The Future of
Sustainability Education at
North American Universities

THE FUTURE OF
SUSTAINABILITY EDUCATION
AT NORTH AMERICAN UNIVERSITIES

Edited by
Naomi Krogman
with
Apryl Bergstrom

UNIVERSITY *of* **ALBERTA** PRESS

Published by

University of Alberta Press
1-16 Rutherford Library South
11204 89 Avenue NW
Edmonton, Alberta, Canada T6G 2J4
amiskwaciwâskahikan | Treaty 6 |
Métis Territory
uap.ualberta.ca | uapress@ualberta.ca

Copyright © 2022 University of Alberta Press

LIBRARY AND ARCHIVES CANADA
CATALOGUING IN PUBLICATION

Title: The future of sustainability education at North American universities / edited by Naomi Krogman with Apryl Bergstrom.
Names: Krogman, Naomi, editor. | Bergstrom, Apryl, editor.
Description: Includes bibliographical references.
Identifiers: Canadiana (print) 20220260664 | Canadiana (ebook) 20220260710 | ISBN 9781772126303 (softcover) | ISBN 9781772126662 (EPUB) | ISBN 9781772126679 (PDF)
Subjects: LCSH: Environmental education—Canada. | LCSH: Environmental education—United States. | LCSH: Sustainable living—Study and teaching (Higher)—Canada. | LCSH: Sustainable living—Study and teaching (Higher)—United States. | LCSH: Sustainable development—Study and teaching (Higher)—Canada. | LCSH: Sustainable development—Study and teaching (Higher)—United States.
Classification: LCC GE90.C3 F88 2022 | DDC 333.7071/171—dc23

First edition, first printing, 2022.
First printed and bound in Canada by Blitzprint, Calgary, Alberta.
Copyediting and proofreading by Angela Pietrobon.

This publication is licensed under a Creative Commons licence, Attribution–Noncommercial–No Derivative Works 4.0 International: see www.creativecommons.org. The text may be reproduced for noncommercial purposes, provided that credit is given to the original author. To obtain permission for uses beyond those outlined in the Creative Commons licence, please contact the University of Alberta Press.

University of Alberta Press supports copyright. Copyright fuels creativity, encourages diverse voices, promotes free speech, and creates a vibrant culture. Thank you for buying an authorized edition of this book and for complying with the copyright laws by not reproducing, scanning, or distributing any part of it in any form without permission. You are supporting writers and allowing University of Alberta Press to continue to publish books for every reader.

University of Alberta Press is committed to protecting our natural environment. As part of our efforts, this book is printed on Enviro Paper: it contains 100% post-consumer recycled fibres and is acid- and chlorine-free.

University of Alberta Press gratefully acknowledges the support received for its publishing program from the Government of Canada, the Canada Council for the Arts, and the Government of Alberta through the Alberta Media Fund.

We dedicate this volume to the memory of Thomas Lovejoy, who attended the contributing authors' workshop that formed the genesis of this book. He shared with us his passion for biological diversity and referred to it as the planet's library of life. We are grateful for his contributions to science and policy, his leadership, and his engagement with lawmakers to help them understand the effects of climate change. He exemplified the work of a sustainability scholar, scientist, teacher, and advocate.

Contents

Foreword XI
THOMAS E. LOVEJOY

Preface XV

Introduction XIX
NAOMI KROGMAN

1 **Charting the Landscape** 1
   An Overview of Sustainability Education in Canadian and US Higher Education
   APRYL BERGSTROM

I | Administrator Point of View

2 **Sustainability Thinking** 35
   A View from the "Dark Side"
   ROGER EPP

3 **Sustainability Scholarship and Education** 45
   Opportunities and Strategies for Success
   CHRISTOPHER G. BOONE

4 **How Trends in Public Higher Education Can Support Sustainability Education and Research** 65
   ROBERT H. JONES

II | Skill Sets or Research Capabilities Needed for Sustainability Education

5 Sustainability Education at US and Canadian Tribal Colleges   77
   Its Goals and Implementations, and the Role of Mathematics
   ROBERT E. MEGGINSON

6 Innovation   101
   Connecting Markets and Money
   VICKY J. SHARPE

7 Sustainability and Decision   119
   THOMAS DIETZ

III | Focusing Sustainability Education on Problem-Based Learning

8 Overcoming the Terrors of the Either/Or   135
   ANN DALE

9 Sustainability Education   147
   A Dance Between Knowledge and Experience
   SHIRLEY M. MALCOM

IV | Cultivating Civic-Mindedness, Deliberative Dialogue, and Pathways toward the Public Good

10 Cultivating Courage in an Increasingly Complex, Divided World   161
    TODDI A. STEELMAN

11 Education for Regeneration   173
    PATRICIA E. (ELLIE) PERKINS

V | Unique Perspectives from Professor and Student

12 Education for Sustainability   189
    An Ecological Citizenship Approach in a Neoliberal Age
    ALLISON F.W. GOEBEL

13 Sustainability Pedagogy   201
    Keeping Up with Millennials and Generation Z
    KOUROSH HOUSHMAND

Conclusion 211
NAOMI KROGMAN

Contributors 227

Foreword

FIVE YEARS AGO, I had the very special experience of participating in a workshop for the contributing authors of *The Future of Sustainability Education at North American Universities* in Banff, Alberta, with a stunning view of the Rocky Mountains as our backdrop. It is a wonderful experience to have those discussions brought back to life in this volume and to see the evolution of excellent pathways to advance sustainability education in this book.

A book on sustainability education does not seem as pioneering now as it did in the context of that workshop five years ago. But no institution of higher education, student, or course can ignore the incredible pressures on our planet and our future. We must take the Glasgow COP as signalling that the time is right at hand—NOW. Among other things, biodiversity loss can no longer be considered independently of the planet's climate system, and we are very fortunate that this book exists.

The Future of Sustainability Education at North American Universities presents a set of interdisciplinary perspectives on sustainability in higher education from students, educators, university administrators, and practitioners from both Canada and the United States. These diverse views describe and discuss the significance of a range of sustainability programs, courses, and pedagogies, and explore a broad array of topics, such as Indigenous perspectives, commons governance, systems thinking, and the challenges and opportunities for change at universities and colleges in the United States and Canada. They offer rich descriptions of what could or should be done to address sustainability more holistically and comprehensively, both within and beyond institutions of higher learning. This collection of writings serves the important purpose of expanding the dialogue on the transdisciplinary and evolving field of sustainability in higher education.

The unique contributions of this book allow the reader to be both outside and within the walls of higher education institutions, and to imagine the dynamic role universities can play in holding up a mirror to existing sustainability challenges and in providing a roadmap for positive change. By hearing from university administrators, professors, thought leaders who work closely with universities, and students, the reader will be able to hear the calls for change from different but complimentary vantage points.

American thought about how to advance sustainability education has been led by institutions like Arizona State University, Clemson University, the American Association for the Advancement of Science, and Duke University. Canadian institutions seem at this point to have benefitted from a greater degree of engagement with Indigenous communities than those south of the border, and this reflection has been of particular benefit to Americans like myself. Over all our deliberations—then, current, and future—floats the eloquent vision of Elinor Ostrom. Ostrom, a 2009 Nobel Peace Prize winner in Economic Sciences, encouraged inter- and transdisciplinary scholarship, such as this, to bridge disciplinary teaching and research with practice, and to work with knowledge holders outside of academia on how to manage resources, protect ecosystems, and support sustainable and equitable livelihoods.

What is very clear, given the undeniable urgency of both the climate change and biodiversity agendas, is how the academy needs to fully integrate sustainability education. I can already see that happening at my own university, George Mason, where we now look for ways to integrate the Sustainable Development Goals in our pedagogy, as well as modify our campus and assist northern Virginia toward greater sustainability.

Last weekend I was at William & Mary (so old it was created by Royal Charter) only to find a student body energized about just these very topics. The sustainability agenda has become quite urgent since we met in that stunning setting in Banff. Several chapters in this book provide guidance on how to engage students in social change processes that advance the sustainability agenda, placing students in communities and alongside policy-makers who tackle a wide range of topics such as watershed management, local food self-reliance, and Indigenous natural resource governance of their long-held traditional lands and seas. Collectively, we face the most urgent environmental justice issue of all time in climate change and biodiversity loss, and students are highly energized by sustainability in action. These are all great signs of interest and commitment to sustainability on campus and beyond.

THOMAS E. LOVEJOY
University Professor of Environmental Science and Policy,
George Mason University
November 2021

Preface

THOMAS HENRY HUXLEY REPORTEDLY SAID, "Patience and tenacity of purpose are worth more than twice their weight in cleverness." While the chapters of this book had a hiatus between the time they were originally penned in 2016 and then revised and updated in 2020, it was our collective patience and tenacity that brought this book to completion. The contributing authors were invited to share their ideas during a two-and-a-half-day workshop in 2016 at the Banff Centre for Arts and Creativity in Alberta, Canada. This workshop was funded by the Social Sciences and Humanities Research Council (SSHRC) and the American Association for the Advancement of Science (AAAS), and had some in-kind support from the Department of Resource Economics and Environmental Sociology at the University of Alberta. Funding from SSHRC, AAAS, and the Faculty of Environment at Simon Fraser University also contributed to the production of this book. Dr. Gary Machlis, University Professor of Environmental Sustainability at Clemson University, South Carolina, and I planned and led the workshop, with essential assistance from Apryl Bergstrom and Claire Doll, two graduate students at the University of Alberta.

The workshop was designed to allow thoughtful exchange and revision of opinions. Our workshop room provided views of autumn splendor in the Rocky Mountains of Banff, and the contributing authors sat around a table and took turns presenting and exchanging ideas about the future of sustainability education. In addition to the authors, there were two special guests: Thomas Lovejoy, who is respectfully called the "Godfather of biodiversity" because he coined the term "biological diversity," and the Honourable Linda Duncan, a long-standing and outspoken member of parliament in Edmonton, Alberta (2008–19) who is passionate about sustainability transitions. I will never forget how Dr. Lovejoy referred to biodiversity as the library of the world; once a unique and treasured book or species is lost, its wisdom, its interdependence, is also lost. Linda Duncan spoke about the burgeoning trends in clean energy jobs and other urban renewal positions, which continue to rapidly grow today as we move toward 70 percent of the world's population living in cities by 2050.

On the first evening, a panel of three university students talked about their experiences of earning degrees in sustainability education, and they shared what was most influential, instructive, and meaningful for them along their educational journeys. I recall that they spoke of transformational experiences that involved the head, hands, and heart. This reaffirmed the importance of encouraging students to collaborate in groups to change things. By doing so, they see the interconnections among the elements of expertise, collective will, trial and error, trusting and respectful relationships, and, incidentally, patience and tenacity. Dr. Joan Greer's art and design students at the University of Alberta loaned us artwork on the theme of climate and environmental change to display in our workshop room. To allow continued conversation and connection, meals were served family style in the workshop room. Finally, unscheduled periods permitted time for reflection or walks or forays outside.

Sustainability education has changed since the contributing authors of this book came together. In 2020, the COVID-19 pandemic prompted universities to accelerate their delivery of remote teaching methods, and this trend has mainstreamed virtual access to classrooms, professor lectures, discussion groups, and course materials, as well as digital options for submitting assignments and exams. While this trend was already underway before the 2020 pandemic, instructors and students have expanded their skill sets to integrate data sets, social media, video, audio, artificial intelligence, and various computing platforms into their course instruction and coursework. The ability of higher education to deliver remote teaching could also mean that some degrees can become more affordable or accessible, thereby enabling more students to earn university credits, certificates, and degrees. Sustainability education has increasingly involved online and in-person hybrid formats, blended learning, accelerated programs, part-time options, and more. These changes could portend, for example, a greater emphasis on the micro-credentials that could be associated with sustainability education in the post-secondary sector, such as by offering certificates in life cycle analysis, climate change policy and practice, or sustainable water management. Several authors, particularly Boone, Jones, and Bergstrom, refer to such trends in this volume.

I wish to thank the contributing authors of this book, who have been so kind and thoughtful to work with. I am grateful to each of them for their willingness to update their chapters. I would like to send a special thank you to Dr. Shirley Malcom for her willingness to revise her chapter in the midst of the public horror and outrage at revealed racist harms in the United States as she held her role as the head of Education and Human Resources Programs at the American Association for the Advancement of Science (AAAS) and supported her community. A huge sense of gratitude lives in my heart for Apryl Bergstrom, for her foundational information chapter in this book and her enormous assistance during all stages of its writing and publication. Apryl was involved in the initial SSHRC grant application, workshop planning and delivery, and final submission of the manuscript. I am grateful to Dr. Gary Machlis for his assistance in developing the SSHRC and AAAS grant proposals, planning and leading the workshop, and offering feedback on the first drafts of the chapters. Mat Buntin, the acquisitions editor at the University of Alberta Press, was particularly helpful as Apryl Bergstrom and I prepared the final manuscript.

As dean of the Faculty of Environment at Simon Fraser University, I am buoyed by the vitality, adaptability, and intelligence of my faculty members and their commitment to teaching sustainability and other courses, now mostly online due to the pandemic. Finally, I wish to thank my family members—my husband, Lee, and my daughters, Eva and Antonia—for their encouragement and support as I worked through the final stages of this book. The final stages took place during a spring and summer of personal and collective (pandemic) challenges. Participating with my young adult children in supporting the Black Lives Matter movement, and going on a mother and daughters' backpacking trip on the Juan de Fuca trail of Vancouver Island, fostered my personal growth and sense of gratitude for the love of those people who stand by me, and for the nature that heals my sorrows and inspires hope and awe of life itself. It is hope that sustains our collective commitment to a sustainable future. May you find what nurtures your hope, and I hope this book will be a part of your pathway.

NAOMI KROGMAN
Dean, Faculty of Environment, Simon Fraser University
May 2021

Introduction

NAOMI KROGMAN

I AM CONTINUALLY REMINDED by students who have long since graduated from university, especially international students and friends long into their careers, of the privilege it is to go to university. It is at universities and colleges that students of all ages can explore the complexity of the world, the values that drive decision making, and the suite of options available to seek solutions, as well as gain the knowledge and skills to improve our social and ecological future. If any field of study can embrace the education needed to inform a better world, it is sustainability education.

Each day, scientists and scholars in all fields unveil new links between human well-being and the earth's systems, whether these are with our climate, politics, food, culture, or, more recently, animal–human infectious disease systems. These links inform our understanding of how deeply connected we are and expand our ability to imagine ways to live more harmoniously within our living systems. This book springs from the hope that our higher education systems will inspire, inform, support, and keep pace with the world of possibility for improvements and solutions that different actors and societies have envisioned to advance sustainability. In this book, the authors describe different conceptual and practical pathways for higher education to teach students about how societies could effectively govern their resources within the earth's means, support sustainable livelihoods, promote human health, expand their capacity to solve problems, demonstrate human creativity and ingenuity, and protect human dignity. Each author uses a rich set of knowledge and experience to provide an informed opinion on the best path forward for post-secondary sustainability education. Some authors draw on more evidence from additional sources than others. The authors each anchor their chapter by describing their positionality, or background, that informs their opinions, and provide their definition of sustainability.

This volume is unlike other books in several ways. Its scope is Canada and the United States, and it focuses on the next decade of changes to sustainability in higher education. While this is not the only book that bridges academic scientific content and a public policy problem—e.g., training and preparing students to address future sustainability challenges—it is unique in that it combines academic and non-academic Canadian and American voices, including the voices of administrators, professors, thought leaders, and two students, to speak to the future of sustainability education. These experts cover a wide gamut of biophysical and socio-cultural sciences, as well as education policy and environmental studies. Contributing authors were invited to answer one or both of the following questions: (1) What are the future directions and opportunities of sustainability education at North American universities? (2) What strategies and tactics can be useful now to encourage progress toward those futures? The book includes discussions on the direction that sustainability education in higher education could take, as well as practical strategies for how to bring about change.

There were 2,361 interdisciplinary environmental, sustainability, and energy degree programs in the United States in 2016 (Vincent et al. 2017, 7), and over 700 interdisciplinary programs in the study of humans and nature in a broad range of disciplines in Canadian post-secondary institutions (Kimantas 2014, 24). Each of these programs must periodically be revised and sometimes renewed or discontinued. My hope is that this book will help guide such changes and inform those who are building new higher education programs in Canada and the US, including the professoriate engaged in sustainability education and administrators, such as department chairpersons, college deans, provosts, and university presidents. This book is also intended for education policy-makers, including government decision makers, board members, and elected officials. Leaders in businesses, governments, and non-governmental organizations who have sustainability responsibilities will benefit from understanding the role of universities in climate change solutions, leadership development, and sustainability innovations. Environmental leaders, including leaders in governments and non-governmental organizations, can use this book to understand how universities are delivering sustainability education and identify opportunities to partner with universities. Leaders of charitable organizations who are involved in philanthropy for sustainability and the environment will find this book useful to help articulate the values and outcomes they want to advance in their commitments to students, faculty, and sustainability education. Finally, leaders of government organizations who hire graduates of the sustainability sciences and humanities may see ways to strengthen government–university partnerships on sustainability science, policy, communication, and practice.

This book offers ideas about how higher education can change our thinking, practices, and relationships to inform and model sustainability. Post-secondary students quickly learn, if they do not know already, that some spheres of influence in their lives are beyond their control, and that there are varying levels of uncertainty about what actions lead to specific outcomes. By learning how these spheres of influence (or systems) are tied to different outcomes for people and the planet, students begin to imagine where they can influence such systems while developing a sense of humility about how complicated change can be.

Generation Z, especially those born from the mid-1990s to the mid-2010s, is media and technology savvy, and its members quickly become aware of social movements around the world where people their age are demanding change. Witness, for example, the #FridaysForFuture, #MeToo, and Black Lives Matter movements, in which youth challenged governments, institutions, and people of privilege to address the institutional inertia that has limited effective actions to reduce carbon emissions, sexual harassment, and racial violence and discrimination. It is at universities that key forms of evidence are generated to inform such social movements and evaluate the effectiveness of different programs for change. It is at universities that critical thinking should be fostered and celebrated, and where ties between social justice and the environment can be deeply discussed.

I now provide a brief summary of each of the chapters in the book. In chapter 1, Apryl Bergstrom provides an overview of trends in sustainability in higher education, discussing how an increasing number of Canadian and US higher education institutions (HEIs) are incorporating sustainability into their institutional frameworks, education, research, campus operations, and outreach and collaboration. Bergstrom begins by outlining how a growing number of HEIs have signed national or international declarations, charters, or initiatives on sustainability in higher education, adopted sustainability plans and policies, and established sustainability offices. Next, she highlights several approaches used by HEIs to incorporate sustainable development (SD) into their formal education systems, including creating new SD courses or programs and reorienting their teaching to foster student engagement and promote more effective learning of SD content. She summarizes some ways that HEIs link research and campus operations to education, such as through campus as a living laboratory initiatives. She then briefly discusses how HEIs contribute to sustainability in community outreach and collaboration, to the benefit of both students and local communities. Bergstrom ends by noting that multiple resources and role models are available to help HEIs more fully integrate sustainability into their systems.

The next three chapters feature senior, high-level administrator perspectives on how universities need to approach sustainability and how they can effectively and creatively do so. Roger Epp, Christopher Boone, and Robert H. Jones offer perspectives based on their experiences as deputy provost, dean, and executive vice president for academic affairs and provost, respectively. Roger Epp starts out this next set of chapters, which speak to how universities, at a high level, can effectively and creatively approach sustainability. Epp, a political scientist by training, has held multiple decanal and higher university leadership positions. In his provocative chapter, "Sustainability Thinking: A View from the 'Dark Side,'" Epp argues that universities need the skill and commitment of senior academic administrators to sustain a serious sustainability agenda. Administrators make connections between curriculum, research, and operations, between the university and the place where it is situated, and between the different forms of knowledge within the academy. Borrowing from Michael Burawoy (2011), Epp holds that a university's work rests on the integrated integrity of four kinds of knowledge: professional or specialist knowledge, policy knowledge, critical knowledge, and public knowledge. Epp argues that a fifth form of knowledge is also required to build a deep and dynamic culture of sustainability thinking and practice in higher education. This form of knowledge is phronesis, which Epp describes as "a practical wisdom shaped by experience and marked by a capacity for judgment, the virtue of knowing how to apply sound principles in complex situations toward some good end." Like Dietz in chapter 7 and Perkins in chapter 11, Epp argues that university educators need to nurture a political wisdom that supports action and an understanding of decision-making processes and structures. Epp concludes that campus operations, place, conversations across "knowledge solitudes," and articulate, principled leadership each matter for building a culture of sustainability thinking and practice in universities.

Christopher Boone, dean of the globally recognized College of Global Futures and professor in the School of Sustainability at Arizona State University, highlights several future opportunities for sustainability scholarship and education, as well as strategies and tactics for going forward. In chapter 3, "Sustainability Scholarship and Education: Opportunities and Strategies for Success," Boone holds that sustainability has reached a critical mass and will continue to be taken on as a formal field of research and study. The caveat is that sustainability scholarship and education must continue to be relevant to external communities, and he offers university practices that will support continued relevance. Boone argues that sustainability programming can benefit from focusing on the five key sustainability competencies, and he describes how these competencies can be cultivated through remote learning course modules, informal learning opportunities, lifelong learning curriculum, and place-based education. Finally, Boone outlines strategies for sustainability education and provides recommendations for how universities can foster "boundary spanning" in their offerings, offer campus as a living lab opportunities, cultivate an ethos of inclusive well-being as part of sustainability, and sharpen students' leadership skills.

In chapter 4, "How Trends in Public Higher Education Can Support Sustainability Education and Research," Robert H. Jones, the provost at Clemson University, South Carolina, contends that four university trends may provide new capacity to build and sustain strong academic programs. Jones first describes changes in economic models to fund university operations, after which he highlights shifts in general education pedagogy and learning approaches. He then discusses how there are fundamental changes in curricula and philosophies affiliated with sustainability-related fields—such as the recent blurring of philosophical differences between various sustainability disciplines—that increase faculty and student desires for sustainability content in higher education. The fourth trend, changing demographics of college-eligible students, means that more students are interested in sustainability, although proactive efforts will be needed to attract more minority participation in the sustainability disciplines. Men, women, and minority groups, he holds, show a strong interest in entrepreneurship and innovation, which may provide an opportunity for recruitment into sustainability initiatives. He concludes that program diversity is growing and, correspondingly, appealing to a more diverse student body. Like Malcom in chapter 9, Jones contends that sustainability programs must resonate with the growing number of students across race, class, gender identity, and other intersectionalities to be relevant for the future.

Following these chapters, three highly accomplished senior researchers and thought leaders offer a set of more specific arguments about the skill sets or research capabilities that need to be further developed in sustainability education. In chapter 5, "Sustainability Education at US and Canadian Tribal Colleges: Its Goals and Implementations, and the Role of Mathematics," Robert Megginson, an Indigenous, multi-award-winning math professor and education advisor and the Arthur F. Thurnau Professor of Mathematics at the University of Michigan, suggests several basic mathematics requirements for degrees in the sustainable management of natural resources. Informed by a survey he conducted of Canadian and US tribal colleges, Megginson's chapter focuses on the role of these colleges in supporting the sustainability goals of the peoples they serve. He points out that Canada has only a few programs that emphasize sustainable resource management, Indigenous environmental stewardship, and reclamation. Megginson's survey suggests that none of the tribal colleges in Canada are specifically described as having a sustainability focus. For US tribal colleges, he found that tribes that hold farming land in common or that manage natural resources are more likely to have sustainable agriculture or natural resource management programs. In general, Megginson argues that sustainability education at these institutions rightly tends to focus on the needs of the tribes served by the colleges. As a mathematics professor who applies math to climate science, Megginson notes that mathematics is "the language of science" and that both the theory and practice of sustainability science make extensive use of mathematics. He concludes by calling for some fundamental math skills, and certain types of advanced math courses, to be taught in what are often high-in-demand sustainable management of natural resources programs at tribal colleges.

The next chapter in this grouping is from the unique voice of Vicky J. Sharpe, who, as the founding president and former CEO of Sustainable Development Technology Canada (SDTC), built an internationally renowned global cleantech fund. A leader in financial investing in sustainability solutions, Sharpe recommends ways that academia can contribute to more effective innovation in sustainable technologies, including providing skill sets for those who enter into finance and policy roles. In chapter 6, "Innovation—Connecting Markets and Money," Sharpe describes the innovation ecosystem and its associated risk factors for sustainability innovations. Sharpe emphasizes that innovation involves more than the more narrowly defined research and development (R&D). Instead, it involves five stages that she calls RD4: research, development, demonstration, deployment, and diffusion into the market. Funds are required at each stage. At the beginning of the innovation chain, academia and industry often provide starting points for R&D. At the other end of the chain, conventional investors and lenders provide the capital required to fund market entry and market volume. Sharpe emphasizes that there is a funding gap at the stages of technology development and demonstration (D&D). She suggests several ways that academia can contribute to more effective innovation, such as by providing testing facilities for larger-scale technologies and helping people who enter finance and policy roles more comprehensively appreciate the importance of carbon and other environmental considerations in capital markets. Sharpe makes several recommendations for better incorporating money and markets into sustainability education. She

Introduction XXV

delivers a strong case for how higher education can play a critically important role in influencing the scale-up and financing of sustainable technologies, such as those related to clean energy.

The next chapter that puts a greater emphasis on specific skill sets needed in sustainability education is by Thomas Dietz, one of the most influential senior social scientists in the US on climate change decision making. Dietz discusses decision making in sustainability and the need in sustainability science research and education to foster interdisciplinarity, discuss and learn about values, link scientific analysis to deliberation, and cope with uncertainty. As an internationally renowned researcher on the drivers of environmental change and human well-being, Dietz calls for specific training that teaches students to connect scientific analysis to democratic and/or deliberative decision making. In chapter 7, "Sustainability and Decision," Dietz contends that sustainability science should not only contribute to fundamental knowledge but also support decision making, which is the core of sustainability. There are several related challenges for universities in teaching about the multiple influences over decision making, such as challenges with fostering interdisciplinarity, incorporating multiple values, and coping with uncertainty. Dietz calls for an "analytic deliberative process" or "research processes that link scientific analysis to deliberation with decision makers and interested and affected parties." He makes the point that technological changes, such as artificial intelligence and biotechnology, will have a huge influence on human beings and the environment. Faculty and students not only need to learn how these technologies can provide sustainability solutions, but also how they can create more problems. Like Steelman in chapter 10, Dietz argues that universities must find ways to engage values. Finally, Dietz's chapter calls for ways to cope with uncertainty in research, learning, and engagement. To address this, universities need to foster social learning and openness to changing one's mind in their instructors and students.

The next two chapters offer perspectives from senior scholars and educational leaders on the importance of focusing on critical, wicked problems in sustainability education. They call for problem-focused and interdisciplinary sustainability education, in which the values of justice and equity are incorporated. Ann Dale, a Trudeau Fellow and director of the School of Environment and Sustainability at Royal

Roads University, and Shirley M. Malcom, head of education for the largest professional science organization in the world, the American Association for the Advancement of Science, both offer suggestions on how to focus sustainability education on problem-based learning. In chapter 8, "Overcoming the Terrors of the Either/Or," Ann Dale reflects on the changes needed in universities to embrace the sustainability imperative, i.e., addressing the pressing problems of the world, such as climate change resilience. Dale argues that the disciplinary structure of the university has changed little since World War II, despite criticisms of the global growth imperative and a move to a carbon neutral economy. Traditional disciplines, she maintains, have failed to embrace the sustainability imperative, forcing them to stay on the periphery. Following Max-Neef, Dale holds that transdisciplinarity is "a different manner of seeing the world" that is more systemic and holistic (Max-Neef 2005, 15). Sustainability education must continue to be innovative to incorporate changes in communication and technologies, such as artificial intelligence and the rise of big data. She encourages sustainability educators to offer an interdisciplinary curriculum that partners with local governments, civil society leaders, and the private sector to address real-world challenges.

In chapter 9, "Sustainability Education: A Dance Between Knowledge and Experience," Shirley M. Malcom similarly invites sustainability educators to ask students, "What is your problem?" and "What do you want to understand?" She argues that this problem-based focus is far more important than asking students to declare their majors. As a leader in shaping STEM (Science, Technology, Engineering, and Math) education at universities across the United States, Malcom calls on higher education to ask students, "Which aspect of this very big concept are you taking on? What is the lens through which you want to view or contribute to sustainability?" A central theme in the chapter is that it is not just about learning how to "do things right," but also how to "do the right thing," which means focusing on the most important problems to address in the world and using the best analytical skill set to do so. A goal for student competency, she argues, involves cultivating depth and breadth, where programs foster empathy and the enthusiasm for other people's disciplines, to the point where students can readily see how to apply specific disciplines to problems. Malcom contends that the most difficult challenge for leaders of higher education institutions is how to build sustainability into the fabric/mission of the institution.

Chapters 10 and 11 turn to how sustainability education should foster civic-mindedness, deliberative dialogue, and pathways toward the public good. Toddi Steelman, Stanback Dean of the Nicholas School of the Environment at Duke University, calls for more confrontive pedagogies to address the most vexing sustainability challenges, and she suggests creating incentives for community-engaged scholarship, developing holistic curricula, and hosting controversial conversations. In "Cultivating Courage in an Increasingly Complex, Divided World," Steelman points to three key societal challenges that undergird the need for sustainability in higher education: (1) a distrust in social institutions, such as government and universities; (2) growing economic divides, which indicate a need to enhance our social equity mission; and (3) credibility gaps, where experts are not trusted. Like Epp, Steelman calls for universities to support "place-based, community-engaged research where salience, legitimacy, and credibility can be established through knowledge co-production and collaboration," and where students are involved in such research. Steelman recommends that universities create institutional incentives for community-engaged scholarship (to help define problems and potential solutions), develop holistic curricula where coursework and experiences are designed to achieve desired learning outcomes, and host controversial conversations on important topics. Skillfully done, these conversations can help people find common ground and build empathy.

The next chapter also squarely addresses the importance of understanding approaches to promoting the collective good in sustainability education. Patricia E. (Ellie) Perkins, professor in the Faculty of Environmental Studies at York University, argues that we need to develop education that transmits skills for personal and collective responsibility, conflict resolution, diversity and equity sensitivity, and intolerance for injustice. Perkins teaches ecological economics, community economic development, and critical interdisciplinary research design. In her chapter "Education for Regeneration," she contends that we need to build education that transmits skills for personal and collective responsibility, conflict resolution, diversity and equity sensitivity, and awareness of nature. Important elements of educational transformation include experiential education, team building, spending time outdoors, and basic science and systems literacy. She maintains that North American universities need to assist in replacing the current globalized economic system with a collective politics that can lead to a regeneration of the earth. Sustainability education, in Perkins' view, must address the injustices that prohibit the inclusivity of those who are most susceptible to *un*sustainable development. Like Dietz in chapter 7, Perkins cites Elinor Ostrom's (1990, 2010a, 2010b, 2014) work on commons governance, which describes conditions that can prevent the "tragedy of the commons" and polycentricity, where students learn how different levels of authority and interdependent elements of a social–ecological system interact to make it stronger and more resilient. Perkins also contends that students need to learn about how people creatively organize to protect each other, such as through cooperatives, activist lifestyles, regional watershed management, and social entrepreneurship.

The next two chapters offer more personalized and contrasting perspectives. Allison Goebel, a Queen's National Scholar and mid-career professor in the School of Environmental Studies at Queen's University, advocates an inquiry-based approach to sustainability education, similar to Dale and Malcom. In chapter 12, "Education for Sustainability: An Ecological Citizenship Approach in a Neoliberal Age," Goebel reflects on the approach her team took to remodel an introductory course in environmental studies around inquiry-based learning. Goebel observes that most major environmental issues have gotten worse since 2003, when her course was created, and that the student body has changed as a result of the information age, intensified materialism, and individualism. A central theme of her revamped course is to challenge individualism and encounter the limits of individual action. Students complete modules to explore an environmental issue by following a largely self-directed, step-by-step process of inquiry-based learning. They develop questions related to their topics, are provided time to "do something about it," and then are asked to reflect on the role of their emotions and on incentives and barriers they face in their efforts to make a positive difference.

Kourosh Houshmand, a multi-award-winning graduate student who recently graduated, writes from a student's perspective, and calls for a pedagogy that keeps pace with real-world applications and that is relevant to millennials and those in Generation Z. Teaching innovation as part of sustainability education is especially important to Houshmand, the youngest contributor to the volume, who was an undergraduate student when he first wrote this chapter. In chapter 13, "Sustainability Pedagogy: Keeping Up with Millennials and Generation Z," Houshmand argues that sustainability education uses an antiquated pedagogy that overemphasizes sustainability as environmentalism and alternatively calls for a pedagogy that includes viewing sustainability through an economic lens. Houshmand maintains that he and his cohort belong to the first generation that learned about sustainability in the K–12 education system. Houshmand contends that millennials and those in Generation Z in particular connect their purchasing power to consumer products that contribute to sustainability, and he calls for more course content on innovation, consumer products, and daily living. In particular, Houshmand argues for more pedagogy on "sustainability as a nuanced tool for innovation, branding, and pricing." He uses three examples that could be used to demonstrate the relevance of sustainability to young adult consumers: Airbnb, cold pressed juice, and farm-to-table as sustainability branding. Educators, using such examples, could observe the trends of sustainability branding and encourage their students to critically engage with the drivers and impacts of sustainability products. As a way forward, Houshmand argues that business programs should integrate sustainability concepts into their curricula and that sustainability should be brought into courses on marketing, technological innovation, and pricing. Finally, Houshmand calls for a nimbler pedagogy that keeps up with real-world applications and that is relevant to the lives of millennials and Generation Z.

My conclusion chapter points to the major themes and highlights the common threads and key contrasts in the opinions presented in the book. Resonant throughout this book is a call for universities to foster a more open discussion on what and how they teach, the intended learning outcomes, and how to stay relevant to students as the issues and connections continue to emerge and change. The conclusion of this chapter addresses the question: "So what? Why does sustainability content in higher education matter?" Join me in deliberating on this question and using the rich material in this book as fodder for your next steps.

References

Burawoy, Michael. 2011. "Redefining the Public University: Global and National Contexts." Chapter 2 in *A Manifesto for the Public University*, edited by John Holmwood, 27–41. London: Bloomsbury Academic.

Kimantas, Janet. 2014. "A\J's 2014 Environmental Education Guide: 112 Canadian Colleges and Universities Have More Than 700 Interdisciplinary Programs in the Study of Humans and Nature." *Alternatives Journal* 40 (4): 24–41.

Max-Neef, Manfred A. 2005. "Foundations of Transdisciplinarity." *Ecological Economics* 53, no. 1 (April): 5–16. https://doi.org/10.1016/j.ecolecon.2005.01.014.

Ostrom, Elinor. 1990. *Governing the Commons: The Evolution of Institutions for Collective Action*. New York: Cambridge University Press.

Ostrom, Elinor. 2010a. "Beyond Markets and States: Polycentric Governance of Complex Economic Systems." *American Economic Review* 100, no. 3 (June): 641–72. https://doi.org/10.1257/aer.100.3.641.

Ostrom, Elinor. 2010b. "Polycentric Systems for Coping with Collective Action and Global Environmental Change." *Global Environmental Change* 20, no. 4 (October): 550–57. https://doi.org/10.1016/j.gloenvcha.2010.07.004.

Ostrom, Elinor. 2014. "A Polycentric Approach for Coping with Climate Change." *Annals of Economics and Finance* 15 (1): 97–134.

Vincent, Shirley, Sumedha Rao, Qiyuan Fu, Katt Gu, Xiao Huang, Kaitlyn Lindaman, Elishiva Mittleman, Kien Nguyen, Rachael Rosenstein, and Young-Jun Suh. 2017. *Scope of Interdisciplinary Environmental, Sustainability, and Energy Baccalaureate and Graduate Education in the United States*. Washington, DC: National Council for Science and the Environment.

1
Charting the Landscape
An Overview of Sustainability Education in Canadian and US Higher Education

APRYL BERGSTROM

HOW CAN SOCIETY ADDRESS wicked social and environmental problems and transition to a more sustainable future? While there is no easy answer to this crucial question, many scholars and practitioners—including the authors of this book—argue that higher education plays a key role in tackling sustainability challenges (Stephens and Graham 2010). This chapter highlights how a growing number of higher education institutions (HEIs) in Canada and the United States acknowledge this role through their policies and practices. It outlines how they bring sustainability into their institutional frameworks by signing declarations or embedding sustainability in institutional policies. It reviews several approaches taken to integrate sustainability in their formal education systems, such as creating sustainability courses and programs and adopting learning approaches to teach sustainability more effectively. The chapter briefly describes how HEIs link sustainability teaching to both research and campus operations through initiatives like campus as a living lab, then outlines some of the ways sustainability is implemented in outreach and collaboration. Finally, it touches on the progress that a few HEIs are making in integrating sustainability throughout their systems.

The information in this chapter was compiled by reviewing select peer-reviewed and grey literature on sustainability in US and Canadian higher education. Due to space limitations and the large volume of literature on this topic, I have focused this chapter on material that sets the reader up to appreciate the larger context of sustainability in HEIs. The choice of topics and the views expressed in this chapter were informed by my master's thesis research on the learning objectives and pedagogies in introductory undergraduate sustainability courses in Canada and the United States. They were also informed by my undergraduate degree in environmental and conservation sciences and my experience in teaching two sustainability-related undergraduate courses as a sessional instructor. My views of sustainability education have been further shaped by my conception of sustainability, which I understand as "meeting the needs of present and future generations while substantially reducing poverty and conserving the planet's life support systems" (PNAS 2021).

Numerous educators, scholars, and practitioners argue that urgent economic, social, cultural, and technological transformations are needed to meet the needs of current and future generations and to pursue the goals of social equity and well-being while keeping human impacts within ecological limits (Schneidewind, Singer-Brodowksi, and Augenstein 2016; WCED 1987). In a well-cited paper, for example, Rockström et al. (2009, 472) estimate that humans have transgressed three of the nine planetary boundaries that define a "safe operating space for humanity"—those of climate change, the rate of biodiversity loss, and "human interference with the nitrogen cycle"—and they warn that this could result in "unacceptable environmental change." Taking interactions among planetary boundaries into account, Lade et al. (2020, 123) estimate that the Earth's system is currently "well outside the safe operating space for human impacts," and they call for urgent action to move Earth's system toward a safe operating space. Many acknowledge that this will be challenging to accomplish, however, since "wicked problems" like climate change and global biodiversity loss are multidimensional, interdependent, value-laden, and unpredictable (Klein 2004), and attempted solutions often result in unforeseen consequences (Lake 2012).

Education has an essential role to play in addressing complex sustainability challenges (Cortese 2003; Escrigas 2016; UNESCO 2002). One of the first international conferences to acknowledge this role was the 1972 United Nations (UN) Stockholm Conference on the Human Environment, which linked environmental concerns to education and other segments of society (Calder and Clugston 2003). This role was reinforced in 1977 with the Tbilisi Declaration, an international declaration that attested that students in all fields need environmental education (Calder and Clugston 2003). Education for sustainable development (ESD) was acknowledged formally for the first time in Agenda 21, which came out of the 1992 Rio Earth Summit (Barth and Michelsen 2013). Chapter 36 in Agenda 21 states that education—including formal education, training, and public awareness initiatives—"is critical for promoting sustainable development and improving the capacity of the people to address environment and development issues" (UNCED 1992, 320). While ESD initially focused on environment and development, it was soon broadened to include cultural and social dimensions (Bell 2016). The prominence of ESD was later reinforced by the UN's Decade of Education for Sustainable Development (2005-14) (UNESCO n.d.) and by the UN's Sustainable Development Goals (SDGs). Target 4.7 of Quality Education, the fourth SDG, explicitly states:

> By 2030, ensure that all learners acquire the knowledge and skills needed to promote sustainable development, including, among others, through education for sustainable development and sustainable lifestyles, human rights, gender equality, promotion of a culture of peace and non-violence, global citizenship and appreciation of cultural diversity and of culture's contribution to sustainable development. (UNDESA n.d.)

Many scholars have emphasized the critical role that higher education in particular plays in addressing sustainability problems (Hart et al. 2015). As Cortese (2003) points out, higher education enjoys academic freedom and can employ a diversity of skills to advance new ideas and conduct experiments in sustainable living. It generates and transfers knowledge to society through service, outreach, and the education of future decision makers (Adomssent and Michelsen 2006). It can engage in the multi-decadal partnerships that are needed to identify, study, and solve sustainability issues (Hart et al. 2015). As Steelman attests in this volume, HEIs can help erode discord and build empathy by facilitating conversations on controversial topics. Finally, higher education has a large potential impact; there were roughly 224 million students in tertiary education[1] worldwide in 2018 (UIS n.d.b). In 2018, an estimated 19.6 million students attended degree granting HEIs in the US (NCES 2019), and 2 million students attended public post-secondary education in Canada (Global Affairs Canada 2018).

Many scholars and educators argue that higher education not only has the ability but also the obligation to help address sustainability issues (Cortese 2003; Escrigas 2016). For example, Escrigas (2016) argues that universities have an intergenerational responsibility that stems from their key role in shaping civilization. Given that most people who advance and manage society's institutions are university educated, the *Report and Declaration of the Presidents Conference* from the 1990 Talloires Conference states that "universities bear profound responsibilities to increase the awareness, knowledge, technologies, and tools to create an environmentally sustainable future" (ULSF 2015).

The Institutional Framework

In the past few decades, HEIs have been increasingly recognizing their responsibility to contribute to sustainability efforts by integrating sustainability into their institutional frameworks. They are developing and signing declarations, charters, and initiatives (DCIs) and other voluntary agreements, joining organizations that support sustainability in higher education (SHE), creating sustainability offices, hiring sustainability officers, and embedding sustainability into their institutional policies (Lozano et al. 2015). Drafted at the end of the Talloires Conference in 1990, the Talloires Declaration was the earliest DCI in which universities recognized their responsibility to environmental sustainability and made a commitment to contribute to it (Wals and Blewitt 2010). The Talloires Declaration includes a ten-point action plan to incorporate environmental and sustainability literacy into college and university research, teaching, outreach, and operations (ULSF n.d.). As of February 2021, there were 519 signatories worldwide, including 173 HEIs in the US and 41 in Canada (ULSF 2021). Since 1990, numerous other national and international DCIs have been created, and a growing number of HEIs have become signatories. In the eleven most widely recognized DCIs,[2] two recurring themes are the focus on sustainability issues—such as societal threats and environmental degradation—and the assertion that HEI leaders have a moral obligation to incorporate and institutionalize sustainable development (SD) (Lozano et al. 2015). The 2017 SDG Accord, for example, recognizes the responsibility that educators play in attaining the SDGs, and calls on HEIs to embed the SDGs into activities throughout their systems (EAUC 2021). UNESCO (2014) suggests that the increased reach of these declarations indicates that sustainability is being incorporated into mainstream thinking on the role of universities in society.

Colleges and universities also join organizations or networks that promote SHE. When HEIs do this, they benefit from and contribute to valuable ideas, guidance, support, and feedback mechanisms that support institutional learning and sustainable practice (Whitney 2016). For example, many HEIs have joined University Leaders for a Sustainable Future (ULSF), Learning for a Sustainable Future (LSF), Second Nature, and the Association for the Advancement of Sustainability in Higher Education (AASHE) (Wright and Elliott 2012). Many HEIs in Canada and the US have joined AASHE in particular, which facilitates the sharing of information and resources through a large annual conference, professional development webinars and workshops, and the Campus Sustainability Hub, an online resource that contains thousands of SHE resources (AASHE n.d.a). More than one thousand HEIs in over three dozen countries also use AASHE's Sustainability Tracking, Assessment & Rating System (STARS™) to measure and report on their sustainability performance (AASHE 2020a). Colleges and universities also join networks that promote the SDGs (Paço 2019). Signatories of the SDG Accord, for instance, can join the SDG Accord Learning Network to discuss the SDGs and share best practices. The Sustainable Development Solutions Network (SDSN) promotes the SDGs and the Paris Agreement on Climate Change (SDSN n.d.). As of May 2021, 42 Canadian and 108 US universities (departments or institutions) had joined their respective national SDSN networks. The University Global Coalition (2021) encourages HEIs to adopt the SDGs as an overarching framework. The THE Impact Rankings assess universities against the United Nations' SDGs (Times Higher Education 2021).

HEIs also incorporate sustainability into their institutional frameworks by creating sustainability offices or hiring sustainability officers. Over half of the US and Canadian HEIs graded by the 2011 College Sustainability Report Card[3] had sustainability offices that focused on achieving campus sustainability goals, and more than three-quarters of those offices had full-time sustainability staff (SEI 2011b). In Canada, 35 percent of the 220 post-secondary institutions examined by Vaughter, Wright, and Herbert (2015) had a sustainability office and/or officer.

An increasing number of US and Canadian HEIs are also including sustainability in their institutional policies and/or strategic plans. In 2011, White (2014) found that only 27 HEIs in the US had a campus sustainability plan; by 2020, Eby (2020) had identified 155 sustainability plans in US HEIs. Additionally, 145 out of 452 US HEIs that submitted a STARS™ report to AASHE as of July 2020 reported having a published sustainability plan (AASHE 2020a).[4] In Canada, half of the 220 accredited post-secondary institutions that Vaughter, Wright, and Herbert (2015) investigated had an environment or sustainability policy, mandate, or plan. Of the strategic plans of the Canadian HEIs that Bieler and McKenzie (2017) analyzed, most (forty-one out of fifty) discussed sustainability in their governance, education, research, campus operations, and/or community outreach. Including sustainability in policies, plans, and mandates indicates that it is an institutional priority. Policies often drive an institution's activities (Vaughter, Wright, and Herbert 2015), and a strategic plan aids in envisioning and communicating its organizational goals (Bieler and McKenzie 2017). Sustainability plans enable HEIs to develop integrated goals, objectives, and strategies that include environmental, economic, and equity considerations, and that encompass a range of issues that affect campus land use, operations, administration, and academics (White 2014).

Despite the progress in incorporating sustainability in their strategic plans, there is room for HEIs to engage more fully with sustainability in their policies and strategic planning. Of the fifty strategic plans of Canadian HEIs analyzed by Bieler and McKenzie (2017), for example, twenty had an "accommodative" response that addressed sustainability in just one or two domains, and twenty-one had a "reformative" or "progressive" response that discussed sustainability in greater detail and addressed it in more domains. No plans included a "transformative" response that addressed sustainability in all domains and that re-thought educational paradigms related to ecology, place, land, and community (Bieler and McKenzie 2017). Additionally, sustainability policies and plans in US and Canadian HEIs have tended to focus more on campus operations than on teaching, learning, and research (Lidstone, Wright, and Sherren 2015; White 2014), which suggests that there is room for those HEIs to more fully incorporate sustainability into education and research policies.

Education

Many North American HEIs are educating for sustainability to some extent (Wright and Elliott 2012), recognizing that graduates need special knowledge, skills, and competencies to tackle complex sustainability issues and contribute to a renewed future (Wiek, Withycombe, and Redman 2011). This section will briefly review some of the terminology used in sustainability education discussions, then outline several approaches to implementing sustainability in formal education. Finally, it will highlight several learning approaches that are often recommended for sustainability education.

Terms that are used in discussions about sustainability education include "sustainability education," "education for sustainability" (EfS), and "education for sustainable development" (ESD) (Sterling 2004, 44). UNESCO favours the term ESD, which it defines as "a learning process (or approach to teaching) based on the ideals and principles that underlie sustainability and...concerned with all levels and types of education" (Wals 2009, 26). For higher education, the terms "sustainability in higher education" (SHE) (Wright 2004) or "higher education for sustainability development" (HESD) (Barth and Rieckmann 2015) are often used.

Sustainability education includes several types of education, such as training and formal, non-formal, and informal education (Buckler and Creech 2014). Formal education occurs in schools, colleges, and universities and is based on an established curriculum and approved teaching methods. Non-formal education takes place in organized learning settings outside of the formal system (e.g., museums). Informal education occurs within families, community groups, media, and other settings (Buckler and Creech 2014). While all forms of education are important in sustainability education, this section will focus on formal education.

Approaches to Implementing Sustainability in Formal Education

Lozano, Ceulemans, and Scarff Seatter (2015, 206) identify five main approaches that universities and colleges use to implement sustainability in formal education: (1) cover some environmental issues and material in existing courses; (2) create SD courses; (3) create SD undergraduate or graduate degree programs; (4) intertwine SD as a concept in disciplinary courses; and (5) offer SD as a specialization within faculty frameworks. This section will briefly describe these approaches, as well as the progress of some HEIs in integrating sustainability across the curriculum.

First, a relatively simple way to incorporate sustainability into formal education is to add environmental or SD material to existing courses (Lozano et al. 2015). In this approach, sustainability is treated as an add-on and is often taught in one or more lectures (Lozano 2010). One criticism of this approach, however, is that students learn material for the course, but then cannot integrate sustainability principles into their professional lives (Lozano et al. 2015).

A second approach is to offer stand-alone, sustainability-focused courses. Many of the Canadian and US HEIs that submit reports to AASHE through the STARS™ reporting system claim to offer a number of sustainability courses and courses that include sustainability.[5] According to Peet, Mulder, and Bijma (2004), sustainability-focused courses may be better suited to addressing sustainability topics that transcend specific disciplinary courses. For example, the topic of how to stimulate sustainable energy supplies is beyond the scope of regular engineering courses, but this could be addressed in a stand-alone sustainability course (Peet, Mulder, and Bijma 2004). Another advantage of stand-alone courses is that they can be developed and implemented more quickly than sustainability programs (Hegarty et al. 2011). However, as Malcom notes in this volume, adding courses around traditional disciplines might not create an integrated learning experience and could lead to an "overstuffed curriculum." As with the criticism of introducing SD content to existing courses, some criticize stand-alone courses for failing to assist students with integrating sustainability concepts in their professional lives (Lozano et al. 2015). Students might therefore view these courses as irrelevant to their disciplines, something merely to be "passed and forgotten" (Haigh 2005, 38).

A growing number of US and Canadian HEIs are offering sustainability degree programs, which provide an opportunity to learn about sustainability in greater depth. According to a periodic census in the US, the number of sustainability degree programs in US four-year universities and colleges increased from one program in 2006 to 141 programs in 2012 (Vincent, Bunn, and Stevens 2013, 5). They increased an additional 89 percent between 2012 and 2016 (Vincent et al. 2017, 7). Although Canada does not conduct a similar census, the types of programs offered by Canadian universities are annually reported in a directory by Universities Canada. According to this directory, the number of sustainability studies undergraduate and graduate degree programs, certificates, or diplomas in English increased from twenty-three in 2016 to thirty-seven in 2020 (Universities Canada 2016, 2020).

More broadly, US HEIs offer a growing number of interdisciplinary environmental, sustainability, and energy (IESE) degree programs in which sustainability is a core principle (Vincent et al. 2017). There were 2,361 IESE degrees in a 2016 census of 2,327 degree-granting institutions and special focus institutions, which is an increase of 15 percent from the 2012 census (Vincent et al. 2017, 6). In Canada, environmental or sustainability programs commonly offered in 2020 included environmental science (77), environmental studies (84), natural resources management and policy, general (42), natural resources/conservation, general (20), sustainability science (37), ecology (27), environmental biology (28), and science, technology, and society (17) (Universities Canada 2020). The numbers in brackets indicate the number of undergraduate and graduate degree programs, certificates, or diplomas in each subject area.

The types of sustainability programs have also expanded in recent years. In 2012, most US sustainability programs fell into the categories of general sustainability, natural resources or environmental sustainability, sustainable development/communities, a combination of sustainability and environmental science or studies, global or international sustainability, and sustainability management (Vincent, Bunn, and Stevens 2013). By 2016, new sustainability programs included sustainability leadership, "social innovation and change for sustainability," sustainable energy, sustainable natural resources and conservation, sustainable cities or communities, and sustainable design (Vincent et al. 2017, 11).

HEIS also intertwine sustainability as a concept in regular disciplinary courses. In this approach, sustainability content is tailored to each course, explicitly stated in course aims, and integrated into teaching and grading (Lozano 2010). This approach is beneficial because it helps institutionalize SD in HEIS (Lozano 2010), provides sustainability education to non-sustainability majors (Lozano et al. 2015), and aids students in integrating sustainability into their disciplines (Peet, Mulder, and Bijma 2004). This is considered important, because if universities fail to teach about sustainability within the context of multiple disciplines, students can internalize the message that sustainability is not relevant for their fields (Hegarty et al. 2011). One challenge of this approach is that many instructors are unfamiliar with SD concepts (Ceulemans and De Prins 2010). Fortunately, there are many resources to assist faculty who want to integrate sustainability into their courses, such as workshops offered by the Piedmont/Ponderosa Model of Faculty Development (Barlett and Chase 2012) and the Disciplinary Associations Network for Sustainability (DANS), which provides discipline-specific resources for teaching sustainability topics (AASHE 2020b).

Many HEIS also offer sustainability as a specialization within each faculty's framework, where students take multiple courses with sustainability content while pursuing a degree in their own discipline. A 2016 census in the US revealed that sustainability concentrations are available in many types of degree programs, such as design and planning, architecture, business management, and engineering (Vincent et al. 2017). More generally, 2,222 degree programs in a wide range of professional fields and disciplines included specializations in IESE in 2016 (Vincent et al. 2017). Specializations such as these allow students in many fields to gain greater insights into sustainability (Kamp 2006).

To create sustainable change, many scholars advocate infusing sustainability in multiple disciplines and integrating it across the entire curriculum (Wright and Elliott 2012), thus exposing students to sustainability throughout their programs (Chiong, Mohamad, and Abdul Aziz 2017). Several US and Canadian HEIS are taking steps to do this. At Fleming College in Ontario, Canada, for example, 89 to 90 percent of the diploma programs embed Fleming's Sustainability Learning Outcome (Fleming College 2020). As Boone describes in this volume, Arizona State University's (ASU) School of Sustainability supports sustainability across the university in multiple ways, such as by cross-listing courses with those in other units. Despite progress in adding sustainability content to courses and programs, it seems that many HEIs have more work to do before sustainability will be fully integrated across the curriculum. STARS™ reports submitted to AASHE up to the year 2014 by gold and silver-ranked HEIs (the highest rankings achieved at that time) indicate that most of these HEIs offered a low proportion of courses with sustainability content compared with their total course offerings (Benton-Short and Merrigan 2016). Fortunately, several resources can support institutions or faculty members seeking to integrate sustainability more fully throughout the curriculum, including DANS and workshops based on the Piedmont/Ponderosa Model (AASHE 2020b; Barlett and Chase 2012). Another resource is AASHE's "Centers for Sustainability Across the Curriculum" program, in which a select number of HEIs provide professional development opportunities or workshops to their own and other institutions (AASHE n.d.b).

Learning Approaches Recommended for Sustainability Education
Worldwide, universities are reorienting their teaching to help learners understand and take part in SD (Wals 2012). They recognize that transmissive teaching approaches, such as lecturing or presenting, may not fully engage students in sustainability challenges (Wals 2012) or sufficiently encourage personal or social change (Sterling 2004). In addition to these traditional approaches, the ESD literature thus recommends multiple alternative learning approaches or pedagogies to foster student learning and engagement. The following learning approaches are highlighted below because they are frequently recommended for ESD in the literature: active and participatory, experiential, multidisciplinary, interdisciplinary, transdisciplinary, problem-based and project-based (PPBL), collaborative, and place-based learning. Competency-based learning and inquiry-based learning are not included here because they are discussed in this volume by Boone and Goebel, respectively. Since these learning approaches are complementary and share a "family resemblance," many scholars and practitioners recommend using a blend of learning approaches and pedagogies rather than a single learning approach (Wals 2012).

Active and Participatory Learning

Active and participatory learning are widely believed by ESD scholars to be core processes that underpin ESD (Tilbury 2011). In active learning, students are actively involved in analyzing, evaluating, or synthesizing information instead of passively absorbing it (King 1993). In participatory learning, learners participate actively in the learning process, often working with others on a task or issue (Wals 2012). Numerous studies affirm that active learning is more effective than lecturing to help students retain information, solve problems, think, change attitudes, and increase their motivation to learn (Levintova and Mueller 2015). Active and participatory learning also aids in learning to clarify values, envision more positive futures, apply learning to real-world contexts, and develop systems thinking and critical thinking abilities—skills often seen as crucial for understanding and addressing sustainability problems (Tilbury 2011). Critical reflective thinking, for example, enables students to deeply examine the origins of unsustainability and identify the biases and hidden assumptions underlying their own opinions, knowledge, and viewpoints (Tilbury 2011).

A literature review of 229 peer-reviewed SHE articles between 2005 and 2018 identified several teaching techniques commonly adopted by HEIs (Menon and Suresh 2020), most of which involve active or participatory learning. The top four techniques include project/problem-based learning, case studies, service learning, and inquiry-based learning. Project/problem-based learning is described below in the project- and problem-based learning section. The case study method assists in bridging theory and practice (Menon and Suresh 2020) and aims to develop problem-solving skills by engaging students with practical real-world challenges (Remington-Doucette et al. 2013). The experiential learning section touches on service learning, and Goebel discusses inquiry-based learning in this volume. Other active and participatory techniques that are often recommended to engage learners in sustainability include internships, site visits, fieldwork (Menon and Suresh 2020), group work or discussions, debates, role play, experiments (Cotton and Winter 2010), critical reading and writing, and outdoor learning (Tilbury 2011). More recently, educational games, e-learning, poetry, drama, pictures, and movies have also been gaining in popularity in ESD (Menon and Suresh 2020).

Experiential Learning

Experiential learning, or "learning by doing," directly engages learners in the phenomena being studied (Domask 2007, 55). Learners have and reflect on direct experiences, form ideas based on these experiences, and then apply their ideas to new experiences (Sipos, Battisti, and Grimm 2008). Students share the responsibility for learning, and instructors act as facilitators who support students in making sense of their experiences (DiConti 2004). Forms of experiential learning include internships, simulations, fieldwork involving primary research (Domask 2007), and community service learning, where students collaborate with community members and engage in services to meet local needs (Sipos, Battisti, and Grimm 2008). Nature-based service learning, for example, could involve planting native trees or monitoring water quality in local water bodies (Hensley 2015). According to Perkins, writing in this volume, engaging students with non-academics and the natural world gets them involved in vital, equity-enhancing problem solving. It also helps them better understand and appreciate the value of this work and build their awareness of positive ways to impact the natural world (Hensley 2015).

Experiential learning has grown in popularity in higher education in recent decades (Domask 2007; Jones, this volume). An example of a program that emphasizes experiential learning is Dalhousie University's Environment, Sustainability and Society (ESS) undergraduate program in Nova Scotia, Canada (Wright 2013). ESS students participate in internships, sustainability-related research projects, community engagement, work experience, and field work (Dalhousie University n.d.).

Multidisciplinary, Interdisciplinary, and Transdisciplinary Learning

Addressing society's complex and multidimensional challenges requires the ability to think holistically, integrate knowledge of human and natural systems, and collaborate across institutional and disciplinary boundaries (Remington-Doucette et al. 2013). In other words, this requires multidisciplinary, interdisciplinary, and transdisciplinary learning. Multidisciplinarity is often understood as the study of a topic using several disciplines; however, the disciplines do not explicitly share their viewpoints, methods, or ways of knowing (Remington-Doucette et al. 2013). Interdisciplinarity involves exploring sustainability issues from several disciplinary angles to come to an integrative perspective on how to resolve or improve the issues (Wals 2012). Transdisciplinarity integrates academic or expert knowledge with the traditional or practical knowledge of non-academic stakeholders (Remington-Doucette et al. 2013) and engages with human values in real-world problem-solving contexts (Evans 2015a). According to Max-Neef (2005), transdisciplinarity entails a more systemic and holistic way of viewing the world. As Dale describes in this volume, transdisciplinary learning also involves skills such as critical thinking and interpersonal, intrapersonal, cognitive, and communication skills. Transdisciplinary learning is also believed to foster mutual learning between students and stakeholders, enabling students to construct their own understanding of issues (Menon and Suresh 2020).

Multiple higher education institutions have created multidisciplinary or interdisciplinary environmental or sustainability programs (Kimantas 2014; Vincent et al. 2017). Some have also created transdisciplinary sustainability programs, such as Brock University's Master of Sustainability program, which introduces students to concepts and research methods used by transdisciplinary researchers (Brock University 2020). In the Sustainability Studies Bachelor of Arts at Colorado Mountain College (CMC), students take multiple disciplinary courses at the lower division level and engage in interdisciplinary work at the upper division level (Evans 2015b). Internships and service-learning activities with organizations and people engaged in sustainability work provide these students with transdisciplinary learning opportunities.

Problem- and Project-Based Learning (PPBL)
Many scholars and practitioners affirm the importance of PPBL in ESD. PPBL is a form of active learning where students are self-directed and instructors serve as learning facilitators. It combines problem-based learning, which focuses on developing a deep and critical understanding of real-world sustainability problems, and project-based learning, where students often work in teams to understand and develop solutions to real-world sustainability problems (Wiek et al. 2014). According to Wiek et al. (2014), PPBL offers transdisciplinary or transacademic[6] work experiences and promotes the development of interpersonal competence, which prepares students to collaborate with diverse experts on real sustainability challenges. According to Dale, writing in this volume,

working on real-world learning opportunities is engaging and can even be transformative.

One example of an HEI that offers many problem-based, project-based, and hybrid PPBL learning opportunities at undergraduate and graduate levels is the School of Sustainability at ASU (Wiek et al. 2014). In one undergraduate capstone project, for instance, ASU students partnered with a local neighbourhood association on a sustainability appraisal focused on homelessness, childhood obesity, and community development. This appraisal prompted the association to apply for funding for a community garden (Wiek et al. 2014). ASU also provides many campus as a living laboratory opportunities to its students (Boone, this volume). Along with fostering skills and providing valuable experiences for students, projects like these can benefit the campus and local community.

Collaborative Learning

Collaborative learning involves working together to construct knowledge (Cranton 1996). It emphasizes process, which includes exchanging information, ideas, and experiences, and coming to a mutually acceptable understanding (Cranton 1996). Educators take on the role of co-learner or participant, rather than expert (Moore 2005). Learners often take part in group projects, case-based learning, or role play activities, which foster skills in communication, planning, negotiation, conflict resolution, delegation (Remington-Doucette and Musgrove 2015), and knowledge synthesis (Brundiers and Wiek 2011), skills that are vital for addressing sustainability problems (Wiek, Withycombe, and Redman 2011).

ASU offers PPBL-focused workshop courses where students interact collaboratively with student team members and outside stakeholders (Wiek et al. 2014). In one workshop course, for example, self-directed student teams worked with local neighbourhoods to create actionable sustainability interventions on walkability, water quality, and health. Students attended neighbourhood association meetings, conducted interviews, and facilitated community workshops, which developed their collaboration skills (Wiek et al. 2014).

Place-Based Learning

In a literature review of peer-reviewed SHE articles, place-based study, fieldwork, and site visits were often preferred teaching methods in science disciplines (Menon and Suresh 2020). Place-based learning uses local places and communities to investigate specific topics and involves examining their unique ecologies and social characteristics (Hensley 2015). It can involve natural history, experiential learning, outdoor education, action research, and service learning (Gruenewald 2003). According to Gruenewald (2003), place-based learning helps students and teachers directly experience, understand, and influence local life, and makes pedagogy more relevant to their lived experience. Hensley (2015) attests that field trips to ecosystems in the local region during a sustainability course nurtured biophilia and the formation of a deeper relationship with the students' natural and social environment. Similarly, Boone, in this volume, notes that Arizona State University uses place-based education to help students identify their sense of place and understand how it can motivate others to care for what matters to them.

"Family Resemblance" among ESD Learning Approaches

Recommended learning approaches for ESD are often complementary and share a "family resemblance" (Wals 2012). They acknowledge that learning involves multiple disciplines and is not just knowledge based. They focus on real issues that affect and engage learners, and they recognize the importance of having quality interactions with others and with the learning environment (Wals 2012). They often also involve similar activities, such as group work, case studies, fieldwork, experiments, real-world problem solving (Cotton and Winter 2010), and engaging learners with their local settings (Gruenewald 2003).

Finally, it is important to note that there is no one-size-fits-all approach to learning in ESD; instead, scholars and practitioners recommend using a blend of learning approaches and pedagogies that is tailored to the learners, learning context, and available resources (Wals 2012).

Research

Sustainability research plays a vital role in efforts toward a more sustainable future (UNESCO 2014). New forms of knowledge are needed to analyze ecological problems, develop economic, social, and technological innovations, and support the world's population within ecological limits (Schneidewind, Singer-Brodowksi, and Augenstein 2016). As "engines of research and innovation," HEIs contribute to the knowledge and solutions needed to address global changes (UNESCO 2014, 124). They implement sustainability in research through publications, technologies, research centres, collaborations across sectors and with other HEIs, and linkages between research and teaching (Lozano et al. 2015). This section will focus on a few ways that HEIs link sustainability research and teaching, including the role of campus as a living laboratory initiatives in linking education, research, and campus operations. See Leal Filho et al. (2021) for a more comprehensive overview of sustainability research in higher education in general, and Wright and Elliott (2012) for a summary of research activities specific to Canada and the US.

Several scholars have noted the importance of integrating research and education in SHE (Dietz, this volume; Yarime et al. 2012). One way to do this is through research-based education, which creates structured opportunities for students to learn through active enquiry and research (Fung 2017). For example, students can complete sustainability research projects and communicate the findings to various audiences. This will increase their awareness of sustainability problems and develop their skills in terms of research, organization, independence, teamwork, and dealing with uncertainty (Ershova et al. 2019). Students can also collaborate with teachers on research projects, making them co-producers of knowledge and shifting the hierarchical dynamic between educators and students (Clark 2018). Research-based education teaches students that research is a process and that knowledge is continuously evolving and context specific, which assists in teaching them the limits of knowledge. This not only engages and prepares students for the workforce but can even transform their thinking (Clark 2018).

Campus as a living lab initiatives have grown in popularity on many campuses in North America (Wright and Elliott 2012). These initiatives use the campus and local community as real-life contexts to integrate teaching, research, and campus operations, and they provide students with learning opportunities that are potentially transformative (Benton-Short and Merrigan 2016). They also provide a "safe-to-fail environment" where students can experiment with solving sustainability problems (Boone, this volume) and contribute to local sustainability (Epp, this volume).

One example of a university with several campus as a living lab initiatives related to sustainability is the University of British Columbia (UBC) in Canada. UBC's Sustainability-in-Residence program, for example, provides opportunities for any student living on campus to collaborate with on- and off-campus partners to organize activities that promote sustainability topics (Teslenko 2019). The Social Ecological Economic Development Studies (SEEDS) program, which was created to advance sustainability and contribute to UBC's international commitments, promotes the campus as a living lab by engaging roughly one thousand faculty, students, staff, and on-campus partners in over one hundred research projects across themes like biodiversity, energy, climate, water, and food.[7] SEEDS also encourages faculty to align their research projects with classroom learning outcomes and develop sustainability-related curricula by tapping into campus networks. In addition to fostering professional skills, SEEDS empowers students and community members "to construct knowledge through engagement and interaction" (Teslenko 2019, 13).

Campus Operations

A growing number of higher education campuses in the US and Canada are incorporating sustainable practices into their campus operations. This often involves "greening" campus operations through initiatives in greenhouse gas reduction (Buckler and Creech 2014), efficient energy and water use (Finlay and Massey 2012), renewable energy production, building development, waste, grounds maintenance, dining services, purchasing, transportation, or other environmental sustainability initiatives (Lozano et al. 2015; Wright and Elliott 2012). As one of many examples of sustainability initiatives in campus operations, Colgate University installed a geothermal exchange system (Colgate University

2017) and announced in April 2019 that it was the first HEI in New York State to achieve carbon neutrality (Colgate University n.d.).

To a lesser extent, HEIs also implement social sustainability initiatives in their campus operations (Wright and Elliott 2012). One example of such an initiative is a longhouse-style facility at the University of Washington (UW) Seattle campus called wəɬəbʔaltxʷ—Intellectual House.[8] This facility provides a gathering and learning space for American Indian and Alaska Native students, staff, and faculty, as well as those from other communities and cultures. It was designed to increase the success of UW's Native American students by preparing them for leadership roles (University of Washington n.d.).

Visible campus sustainability projects not only reduce the environmental footprint of universities and colleges but, as mentioned previously, can also contribute to education and research through initiatives like campus as a living lab. Additionally, by modelling sustainable design and action (Vaughter, Wright, and Herbert 2015) and visually communicating the values of the HEI (Hopkinson, Hughes, and Layer 2008), they contribute to the "hidden curriculum" that is learned alongside the formal curriculum (Evans 2015a) and reinforce the message that sustainability is important.

Outreach and Collaboration

HEIs also contribute to sustainability through outreach and collaboration (Lozano et al. 2015). Community outreach on sustainability could involve engaging with surrounding communities in mutually beneficial processes of sustainable development, such as creating university-organized sustainability events (Lozano et al. 2015) or short courses available to the community (Berchin et al. 2019). Outreach initiatives can raise awareness, promote community development and resilience, and encourage students to get involved in local communities, which stimulates volunteerism and improves learning via participative processes (Berchin et al. 2019).

Collaborative initiatives can include student exchange programs, joint research, degrees with other HEIs, and SD partnerships with HEIs or non-academic stakeholders (Lozano et al. 2015). The Oberlin Project in Ohio is an example of a partnership to plan and develop a sustainable and prosperous community. Oberlin College, the Oberlin City School District, and the City of Oberlin collaborated to develop a thirteen-acre Green Arts District, shift Oberlin City and Oberlin College to renewable energy sources, and create an alliance of local colleges and schools (Oberlin Project n.d.). Oberlin College and Oberlin City also collaborated to engage students in project-based learning, internships, and research involving community sustainability issues (Daneri, Trencher, and Petersen 2015). Initiatives such as these benefit the community while promoting the development of new skills and knowledge. Collaborations like the Oberlin Project also suggest a shift in how the social function of the academy is perceived, from that of contributing to societal and economic development through technology transfer to that of transforming and co-creating society while working toward sustainable development (Trencher et al. 2014).

Integrating Sustainability throughout the University System

To prepare students to participate in the sustainability transition, Lozano et al. (2013, 11) argue that HEIs should integrate sustainability as a "Golden Thread" throughout their systems. Although HEIs worldwide have begun implementing sustainability in various system components, Lozano et al. (2015) point out that much of this work has been compartmentalized. Similarly, Vincent and Mulkey (2015) attest that few HEIs in the US address sustainability throughout their systems. In Canada, while the policies of many post-secondary institutes acknowledge the importance of embedding sustainability principles across their domains, information about how to achieve associated objectives is often vaguely stated (Vaughter et al. 2016). This suggests that additional work is needed for many HEIs to go beyond aspirational language to concretely implement sustainability across all domains.

Colleges and universities that strive to integrate sustainability more holistically throughout their systems can look to several role models, such as ASU, Unity College, and Cornell University. As Boone describes in this volume, sustainability permeates all aspects of ASU's system, guiding its efforts in education, research, community engagement, and other domains. Unity College, a small environmental college in Maine, reorganized its curriculum, pedagogy, and administrative structure according to a transdisciplinary sustainability science framework (Vincent and Mulkey 2015). Unity is integrating sustainability into its degree offerings and general education requirements and has implemented a transdisciplinary pedagogy in its core curriculum. Instead of departments, it offers degrees through five Centers of Academic Excellence, a structure that enables faculty to share resources and knowledge (Vincent and Mulkey 2015). Cornell University—which earned the highest rating (platinum) in AASHE's STARS™ system for its sustainability initiatives (AASHE 2021)—has integrated sustainability in multiple system domains. For example, it supports environmentally, economically, and socially sustainable leadership in its annually updated report and strategic plan, *Sustainability: Today and Tomorrow* (Salvioni, Franzoni, and Cassano 2017). In addition, 16 percent of its courses are sustainability course offerings, 33 percent of its faculty are engaged in sustainability research (AASHE 2021), and it offers thirty-seven majors and minors focused on sustainability (Cornell University n.d.). More than twenty-five buildings are LEED-certified, seventy-eight green offices and labs are certified, and 20 percent of its electricity comes from renewables (Cornell University n.d.). Finally, the Engaged Cornell initiative has supported over one thousand sustainability-focused projects to promote community engagement in teaching, learning, and research (AASHE 2021).

Conclusion

This chapter has reviewed how Canadian and US HEIs are acknowledging their role and commitment to contribute to a more sustainable future by incorporating sustainability in multiple system domains. They bring sustainability into their institutional frameworks by signing declarations, connecting with organizations, creating institutional policies and sustainability offices, and hiring sustainability officers. They integrate sustainability in formal education by adding SD content to courses, creating SD courses or degree programs, intertwining SD in disciplinary courses, and offering SD as a specialization in each faculty's framework. Some HEIs are also working to integrate sustainability across the entire curriculum. HEIs are adopting recommended learning approaches for ESD, such as active and participatory, experiential, multidisciplinary, interdisciplinary, transdisciplinary, PPBL, collaborative, and place-based learning. Colleges and universities are linking research, education, and campus operations through initiatives like campus as a living lab. Many HEIs are making their campus operations more sustainable, which reinforces the message that sustainability is important. Outreach and collaboration initiatives—such as short courses, sustainability events, or partnerships with HEIs or non-academic stakeholders—foster student learning while promoting community development and resilience.

Finally, there are multiple resources available to assist HEIs with incorporating sustainability in their systems, including organizations like AASHE and programs such as AASHE's "Centers for Sustainability Across the Curriculum." HEIs can also learn from the policies, practices, successes, and failures of other universities and colleges that have integrated sustainability throughout their systems, such as ASU, Unity College, and Cornell University. If HEIs use these and other available resources to weave sustainability as a "golden thread" throughout their systems, they will be better positioned to help society tackle wicked problems and transition to a more sustainable future.

Notes

1. In the UNESCO Institute for Statistics (UIS) database, student enrollment in tertiary education includes those in short-cycle tertiary education (which is often designed to provide professional knowledge, skills, and competencies), and in bachelor's, master's, and doctoral tertiary education or their equivalent levels (UIS n.d.a.).
2. According to Lozano et al. (2015), the eleven most widely-recognized SHE DCIs are: the Talloires Declaration (1990), the Halifax Declaration (1991), the Swansea Declaration (1993), the Kyoto Declaration on Sustainable Development (1993), the COPERNICUS Charter (1994), the Global Higher Education for Sustainability Partnership (GHESP) (2000), the Lüneburg Declaration (2001), the Declaration of Barcelona (2004), the Graz Declaration (2005), the Abuja Declaration (2009), and the Rio+20 Higher Education Sustainability Initiative (2012).
3. From 2007 to 2011, the College Sustainability Report Card created annual sustainability report cards on the HEIs with the three hundred largest endowments in the US and Canada, as well as twenty-two other HEIs that had asked to be included (SEI 2011a).
4. This tally does not include HEIs that have a sustainability plan but did not submit a STARS™ report.

5. Many lists of sustainability courses in the STARS™ reporting system also use the terms "sustainability-focused" or "sustainability-related" courses, which were used in earlier versions of STARS™.
6. In transacademic settings, students contribute to collective problem-solving processes that integrate knowledge from different sectors (Brundiers and Wiek 2011).
7. Project research reports are searchable by these and other themes in the SEEDS Sustainability Library at https://sustain.ubc.ca/teaching-applied-learning/seeds-sustainability-program.
8. wəɬəbʔaltxʷ is from the Luhshootseed language and is pronounced "wah-sheb-altuh" (University of Washington n.d.).

References

AASHE (Association for the Advancement of Sustainability in Higher Education). 2020a. "STARS Participants & Reports." https://reports.aashe.org/institutions/participants-and-reports/.

AASHE (Association for the Advancement of Sustainability in Higher Education). 2020b. "DANS." https://www.aashe.org/partners/dans/.

AASHE (Association for the Advancement of Sustainability in Higher Education). 2021. "Cornell University." https://reports.aashe.org/institutions/cornell-university-ny/report/2021-03-04/.

AASHE (Association for the Advancement of Sustainability in Higher Education). n.d.a. "About the Campus Sustainability Hub." https://hub.aashe.org/about/.

AASHE (Association for the Advancement of Sustainability in Higher Education). n.d.b. "Centers for Sustainability across the Curriculum." https://www.aashe.org/partners/centers-for-sustainability-across-the-curriculum/.

Adomssent, Maik, and Gerd Michelsen. 2006. "German Academia Heading for Sustainability? Reflections on Policy and Practice in Teaching, Research and Institutional Innovations." *Environmental Education Research* 12, no. 1 (February): 85–99. https://doi.org/10.1080/13504620500527758.

Barlett, Peggy F., and Geoffrey W. Chase. 2012. "Curricular Innovation for Sustainability: The Piedmont/Ponderosa Model of Faculty Development." *Liberal Education* 98, no. 4 (Fall): 14–21.

Barth, Matthias, and Gerd Michelsen. 2013. "Learning for Change: An Educational Contribution to Sustainability Science." *Sustainability Science* 8 :103–19. https://doi.org/10.1007/s11625-012-0181-5.

Barth, Matthias, and Marco Rieckmann. 2015. "State of the Art in Research on Higher Education for Sustainable Development." In *Routledge Handbook of Higher Education for Sustainable Development*, edited by Matthias Barth, Gerd Michelsen, Marco Rieckmann, and Ian Thomas, 100–13. London: Routledge.

Bell, David V.J. 2016. "Twenty-First Century Education: Transformative Education for Sustainability and Responsible Citizenship." *Journal of Teacher Education for Sustainability* 18 (1): 48–56.

Benton-Short, Lisa, and Kathleen A. Merrigan. 2016. "Beyond Interdisciplinary: How Sustainability Creates Opportunities for Pan-University Efforts." *Journal of Environmental Studies and Sciences* 6, no. 2 (June): 387–98. https://doi.org/10.1007/s13412-015-0341-x.

Berchin, Issa Ibrahim, Stephane Louise Boca Santa, and José Baltazar Salgueirinho Osório de Andrade Guerra. 2019. "Community Outreach on Sustainability." In *Encyclopedia of Sustainability in Higher Education*, edited by Walter Leal Filho, 250–54. Cham: Springer.

Bieler, Andrew, and Marcia McKenzie. 2017. "Strategic Planning for Sustainability in Canadian Higher Education." *Sustainability* 9, no. 2 (February): 161. https://doi.org/10.3390/su9020161.

Brock University. 2020. "Sustainability Science and Society." 2020–2021 Graduate Calendar. https://brocku.ca/webcal/2020/graduate/snss.html.

Brundiers, Katja, and Arnim Wiek. 2011. "Educating Students in Real-World Sustainability Research: Vision and Implementation." *Innovative Higher Education* 36, no. 2 (April): 107–24. https://doi.org/10.1007/s10755-010-9161-9.

Buckler, Carolee, and Heather Creech. 2014. *Shaping the Future We Want: UN Decade of Education for Sustainable Development (2005–2014) Final Report.* Paris: UNESCO. https://unesdoc.unesco.org/ark:/48223/pf0000230171.locale=en.

Calder, Wynn, and Richard M. Clugston. 2003. "Progress toward Sustainability in Higher Education." *Environmental Law Reporter News & Analysis* 33, no. 1 (January): 10003–23.

Ceulemans, Kim, and Marijke De Prins. 2010. "Teacher's Manual and Method for SD Integration in Curricula." *Journal of Cleaner Production* 18, no. 7 (May): 645–51. https://doi.org/10.1016/j.jclepro.2009.09.014.

Chiong, K.S., Z.F. Mohamad, and A.R. Abdul Aziz. 2017. "Factors Encouraging Sustainability Integration into Institutions of Higher Education." *International Journal of Environmental Science and Technology* 14, no. 4 (April): 911–22. https://doi.org/10.1007/s13762-016-1164-3.

Clark, Lauren. 2018. "Research-Based Education: Engaging Staff and Students in Praxis." In *Shaping Higher Education with Students: Ways to Connect Research and Teaching*, edited by Vincent C.H. Tong, Alex Standen, and Mina Sotiriou, 73–79. London: UCL Press.

Colgate University. 2017. "Sustainability Showcase: Chapel House." *Sustainability News*, January 27, 2017. http://blogs.colgate.edu/sustainability/2017/01/27/sustainability-showcase-chapel-house/.

Colgate University. n.d. "Carbon Neutrality." Accessed July 17, 2020. https://www.colgate.edu/about/sustainability/carbon-neutrality.

Cornell University. n.d. "Sustainability." Accessed May 26, 2021. https://sustainability.cornell.edu/.

Cortese, Anthony D. 2003. "The Critical Role of Higher Education in Creating a Sustainable Future." *Planning for Higher Education* 31, no. 3 (March–May): 15–22.

Cotton, Debby, and Jennie Winter. 2010. "'It's Not Just Bits of Paper and Light Bulbs': A Review of Sustainability Pedagogies and Their Potential for Use in Higher Education." In *Sustainability Education: Perspectives and Practice across Higher Education*, edited by Paula Jones, David Selby, and Stephen Sterling, 39–54. New York: Earthscan.

Cranton, Patricia. 1996. "Types of Group Learning." *New Directions for Adult and Continuing Education* 1996, no. 71 (Fall): 25–32. https://doi.org/10.1002/ace.36719967105.

Dalhousie University. n.d. "Why Study Environment, Sustainability and Society at Dal?" Accessed July 15, 2020. https://www.dal.ca/academics/programs/undergraduate/ess.html.

Daneri, Daniel Rosenberg, Gregory Trencher, and John Petersen. 2015. "Students as Change Agents in a Town-Wide Sustainability Transformation: The Oberlin Project at Oberlin College." *Current Opinion in Environmental Sustainability* 16 (October): 14–21. https://doi.org/10.1016/j.cosust.2015.07.005.

DiConti, Veronica Donahue. 2004. "Experiential Education in a Knowledge-Based Economy: Is it Time to Reexamine the Liberal Arts?" *The Journal of General Education* 53 (3–4): 167–83. https://doi.org/10.1353/jge.2005.0003.

Domask, Joseph J. 2007. "Achieving Goals in Higher Education: An Experiential Approach to Sustainability Studies." *International Journal of Sustainability in Higher Education* 8 (1): 53–68. https://doi.org/10.1108/14676370710717599.

EAUC (The Alliance for Sustainability Leadership in Education). 2021. "The SDG Accord." https://www.eauc.org.uk/the_sdg_accord.

Eby, Robert Freeman. 2020. "Campus Sustainability Plans: A Descriptive Analysis of Sustainability Plans from Institutions of Higher Education in the United States." Unpublished Master of Science Thesis, Texas State University. https://digital.library.txstate.edu/handle/10877/12251.

Ershova, Alexandra, Tatiana Eremina, Mikhail Shilin, and Olga Khaimina. 2019. "Research-Based Teaching Methods for Sustainable Development." In *Encyclopedia of Sustainability in Higher Education*, edited by Walter Leal Filho, 1393–98. Cham: Springer.

Escrigas, Cristina. 2016. "A Higher Calling for Higher Education." *Great Transition Initiative: Toward a Transformative Vision and Praxis*. http://www.greattransition.org/publication/a-higher-calling-for-higher-education.

Evans, Tina L. 2015a. "Transdisciplinary Collaborations for Sustainability Education: Institutional and Intragroup Challenges and Opportunities." *Policy Futures in Education* 13, no. 1 (January): 70–96. https://doi.org/10.1177/1478210314566731.

Evans, Tina L. 2015b. "Finding Heart: Generating and Maintaining Hope and Agency through Sustainability Education." *Journal of Sustainability Education* 10 (November).

Finlay, Jessica, and Jennifer Massey. 2012. "Eco-Campus: Applying the Eco-City Model to Develop Green University and College Campuses." *International Journal of Sustainability in Higher Education* 13 (2): 150–65. https://doi.org/10.1108/14676371211211836.

Fleming College. 2020. "Office of Sustainability: Sustainability in Our Courses." https://flemingcollege.ca/sustainability/sustainability-in-our-courses.

Fung, Dilly. 2017. *A Connected Curriculum for Higher Education*. London: UCL Press.

Global Affairs Canada. 2018. "Canada's Implementation of the 2030 Agenda for Sustainable Development: Voluntary National Review." https://publications.gc.ca/site/eng/9.858493/publication.html.

Gruenewald, David A. 2003. "Foundations of Place: A Multidisciplinary Framework for Place-Conscious Education." *American Educational Research Journal* 40, no. 3 (Fall): 619–54. https://doi.org/10.3102/00028312040003619.

Haigh, Martin. 2005. "Greening the University Curriculum: Appraising an International Movement." *Journal of Geography in Higher Education* 29 (1): 31–48. https://doi.org/10.1080/03098260500030355.

Hart, David D., Kathleen P. Bell, Laura A. Lindenfeld, Shaleen Jain, Teresa R. Johnson, Darren Ranco, and Brian McGill. 2015. "Strengthening the Role of Universities in Addressing Sustainability Challenges: The Mitchell Center for Sustainability Solutions as an Institutional Experiment." *Ecology and Society* 20 (2): 4. https://doi.org/10.5751/ES-07283-200204.

Hegarty, Kathryn, Ian Thomas, Cathryn Kriewaldt, Sarah Holdsworth, and Sarah Bekessy. 2011. "Insights into the Value of a 'Stand-Alone' Course for Sustainability Education." *Environmental Education Research* 17 (4): 451–69. https://doi.org/10.1080/13504622.2010.547931.

Hensley, Nathan. 2015. "Cultivating Biophilia: Utilizing Direct Experience to Promote Environmental Sustainability." *Journal of Sustainability Education* 9 (March). http://www.susted.com/wordpress/content/cultivating-biophilia-utilizing-direct-experience-to-promote-environmental-sustainability_2015_03/.

Hopkinson, Peter, Peter Hughes, and Geoff Layer. 2008. "Sustainable Graduates: Linking Formal, Informal and Campus Curricula to Embed Education for Sustainable Development in the Student Learning Experience." *Environmental Education Research* 14 (4): 435–54. https://doi.org/10.1080/13504620802283100.

Kamp, Linda. 2006. "Engineering Education in Sustainable Development at Delft University of Technology." *Journal of Cleaner Production* 14 (9–11): 928–31. https://doi.org/10.1016/j.jclepro.2005.11.036.

Kimantas, Janet. 2014. "A\J's 2014 Environmental Education Guide: 112 Canadian Colleges and Universities Have More Than 700 Interdisciplinary Programs in the Study of Humans and Nature." *Alternatives Journal* 40 (4): 24–41.

King, Alison. 1993. "From Sage on the Stage to Guide on the Side." *College Teaching* 41 (1): 30–35. https://doi.org/10.1080/87567555.1993.9926781.

Klein, Julie T. 2004. "Interdisciplinarity and Complexity: An Evolving Relationship." *Emergence: Complexity and Organization (E:CO)* 6, no. 1–2 (Fall): 2–10.

Lade, Steven J., Will Steffen, Wim De Vries, Stephen R. Carpenter, Jonathan F. Donges, Dieter Gerten, Holger Hoff, Tim Newbold, Katherine Richardson, and Johan Rockström. 2020. "Human Impacts on Planetary Boundaries Amplified by Earth System Interactions." *Nature Sustainability* 3, no. 2 (February): 119–28. https://doi.org/10.1038/s41893-019-0454-4.

Lake, Danielle. 2012. "Sustainability as a Core Issue in Diversity and Critical Thinking Education." In *Teaching Sustainability/Teaching Sustainably*, edited by Kirsten Allen Bartels and Kelly A. Parker, 31–40. Sterling, VA: Stylus Publishing, LLC.

Leal Filho, Walter, Markus Will, Chris Shiel, Arminda Paço, Carla Sofia Farinha, Violeta Orlovic Lovren, Lucas Veiga Avila, Johannes (Joost) Platje, Ayyoob Sharifi, Claudio R.P. Vasconcelos, Barbara Maria Fritzen Gomes, Amanda Lange Salvia, Rosley Anholon, Izabella Rampasso, Osvaldo L.G. Quelhas, and Antonis Skouloudis. 2021. "Towards a Common Future: Revising the Evolution of University-Based Sustainability Research Literature." *International Journal of Sustainable Development & World Ecology* 28 (6): 503–17. https://doi.org/10.1080/13504509.2021.1881651.

Levintova, Ekaterina M., and David W. Mueller. 2015. "Sustainability: Teaching an Interdisciplinary Threshold Concept through Traditional Lecture and Active Learning." *The Canadian Journal for the Scholarship of Teaching and Learning* 6, no. 1 (article 3). https://doi.org/10.5206/cjsotl-rcacea.2015.1.3.

Lidstone, Lauri, Tarah Wright, and Kate Sherren. 2015. "An Analysis of Canadian STARS-Rated Higher Education Sustainability Policies." *Environment, Development and Sustainability* 17, no. 2 (April): 259–78. https://doi.org/10.1007/s10668-014-9598-6.

Lozano, Rodrigo. 2010. "Diffusion of Sustainable Development in Universities' Curricula: An Empirical Example from Cardiff University." *Journal of Cleaner Production* 18, no. 7 (May): 637–44. https://doi.org/10.1016/j.jclepro.2009.07.005.

Lozano, Rodrigo, Kim Ceulemans, and Carol Scarff Seatter. 2015. "Teaching Organisational Change Management for Sustainability: Designing and Delivering a Course at the University of Leeds to Better Prepare Future Sustainability Change Agents." *Journal of Cleaner Production* 106 (November): 205–15.

Lozano, Rodrigo, Kim Ceulemans, Mar Alonso-Almeida, Donald Huisingh, Francisco J. Lozano, Tom Waas, Wim Lambrechts, Rebeka Lukman, and Jean Hugé. 2015. "A Review of Commitment and Implementation of Sustainable Development in Higher Education: Results from a Worldwide Survey." *Journal of Cleaner Production* 108, Part A (December): 1–18. https://doi.org/10.1016/j.jclepro.2014.09.048.

Lozano, Rodrigo, Rebeka Lukman, Francisco J. Lozano, Donald Huisingh, and Wim Lambrechts. 2013. "Declarations for Sustainability in Higher Education: Becoming Better Leaders, through Addressing the University System." *Journal of Cleaner Production* 48 (June): 10–19. https://doi.org/10.1016/j.jclepro.2011.10.006.

Max-Neef, Manfred A. 2005. "Foundations of Transdisciplinarity." *Ecological Economics* 53, no. 1 (April): 5–16. https://doi.org/10.1016/j.ecolecon.2005.01.014.

Menon, Shalini, and M. Suresh. 2020. "Synergizing Education, Research, Campus Operations, and Community Engagements towards Sustainability in Higher Education: A Literature Review." *International Journal of Sustainability in Higher Education* 21 (5): 1015–51. https://doi.org/10.1108/IJSHE-03-2020-0089.

Moore, Janet. 2005. "Is Higher Education Ready for Transformative Learning? A Question Explored in the Study of Sustainability." *Journal of Transformative Education* 3, no. 1 (January): 76–91. https://doi.org/10.1177/1541344604270862.

NCES (National Center for Education Statistics). 2019. "Table 306.10. Total Fall Enrollment in Degree-Granting Postsecondary Institutions, by Level of Enrollment, Sex, Attendance Status, and Race/Ethnicity or Nonresident Alien Status of Student: Selected Years, 1976 through 2018." Digest of Education Statistics. https://nces.ed.gov/programs/digest/d19/tables/dt19_306.10.asp.

Oberlin Project. n.d. "About." Accessed July 16, 2020. https://oberlinproject.org/about/.

Paço, Arminda. 2019. "Sustainable Development Goals." In *Encyclopedia of Sustainability in Higher Education*, edited by Walter Leal Filho, 1766–71. Cham: Springer.

Peet, D.-J., Karel F. Mulder, and Arianne Bijma. 2004. "Integrating SD into Engineering Courses at the Delft University of Technology: The Individual Interaction Method." *International Journal of Sustainability in Higher Education* 5, no. 3 (September): 278–88. https://doi.org/10.1108/14676370410546420.

PNAS (Proceedings of the National Academy of Sciences of the United States of America). 2021. "Sustainability Science." https://www.pnas.org/portal/sustainability.

Remington-Doucette, Sonya M., Kim Y. Hiller Connell, Cosette M. Armstrong, and Sheryl L. Musgrove. 2013. "Assessing Sustainability Education in a Transdisciplinary Undergraduate Course Focused on Real-World Problem Solving: A Case for Disciplinary Grounding." *International Journal of Sustainability in Higher Education* 14 (4): 404–33. https://doi.org/10.1108/IJSHE-01-2012-0001.

Remington-Doucette, Sonya M., and Sheryl Musgrove. 2015. "Variation in Sustainability Competency Development According to Age, Gender, and Disciplinary Affiliation: Implications for Teaching Practice and Overall Program Structure." *International Journal of Sustainability in Higher Education* 16 (4): 537–75. https://doi.org/10.1108/IJSHE-01-2013-0005.

Rockström Johan, Will Steffen, Kevin Noone, Åsa Persson, F. Stuart Chapin, III, Eric F. Lambin, Timothy M. Lenton, Marten Scheffer, Carl Folke, Hans Joachim Schellnhuber, Björn Nykvist, Cynthia A. de Wit, Terry Hughes, Sander van der Leeuw, Henning Rodhe, Sverker Sörlin, Peter K. Snyder, Robert Costanza, Uno Svedin, Malin Falkenmark, Louise Karlberg, Robert W. Corell, Victoria J. Fabry, James Hansen, Brian Walker, Diana Liverman, Katherine Richardson, Paul Crutzen, and Jonathan A. Foley. 2009. "A Safe Operating Space for Humanity." *Nature* 461, no. 7263 (September): 472–75. https://doi.org/10.1038/461472a.

Salvioni, Daniela M., Simona Franzoni, and Raffaella Cassano. 2017. "Sustainability in the Higher Education System: An Opportunity to Improve Quality and Image." *Sustainability* 9 (6): 914. https://doi.org/10.3390/su9060914.

Schneidewind, Uwe, Mandy Singer-Brodowksi, and Karoline Augenstein. 2016. "Sustainability and Science Policy." In *Sustainability Science: An Introduction*, edited by Harald Heinrichs, Pim Martens, Gerd Michelsen, and Arnim Wiek, 149–60. Dordrecht: Springer.

SDSN (Sustainable Development Solutions Network). n.d. "About Us." Accessed May 3, 2021. https://www.unsdsn.org/about-us.

SEI (Sustainable Endowments Institute). 2011a. "Report Card 2011: Executive Summary." The College Sustainability Report Card. Accessed May 15, 2021. http://www.greenreportcard.org/report-card-2011/executive-summary.html.

SEI (Sustainable Endowments Institute). 2011b. "Report Card 2011: Administration." The College Sustainability Report Card. Accessed May 15, 2021. http://www.greenreportcard.org/report-card-2011/categories/administration.html.

Sipos, Yona, Bryce Battisti, and Kurt Grimm. 2008. "Achieving Transformative Sustainability Learning: Engaging Heads, Hands and Heart." *International Journal of Sustainability in Higher Education* 9 (1): 68–86. https://doi.org/10.1108/14676370810842193.

Stephens, Jennie C., and Amanda C. Graham. 2010. "Toward an Empirical Research Agenda for Sustainability in Higher Education: Exploring the Transition Management Framework." *Journal of Cleaner Production* 18, no. 7 (May): 611–18. https://doi.org/10.1016/j.jclepro.2009.07.009.

Sterling, Stephen. 2004. "An Analysis of the Development of Sustainability Education Internationally: Evolution, Interpretation and Transformative Potential." In *The Sustainability Curriculum: The Challenge for Higher Education*, edited by John Blewitt and Cedric Cullingford, 43–62. London: Earthscan.

Teslenko, Tatiana. 2019. "Engaging Students and Campus Community in Sustainability Activities in a Major Canadian University." In *Sustainability on University Campuses: Learning, Skills Building and Best Practices*, edited by Walter Leal Filho and Ugo Bardi, 3–20. World Sustainability Series. Cham: Springer.

Tilbury, Daniella. 2011. *Education for Sustainable Development: An Expert Review of Processes and Learning*. Paris: UNESCO. https://unesdoc.unesco.org/ark:/48223/pf0000191442.

Times Higher Education. 2021. "Impact Rankings 2021." https://www.timeshighereducation.com/impactrankings#!/page/0/length/25/sort_by/rank/sort_order/asc/cols/undefined.

Trencher, Gregory, Masaru Yarime, Kes B. McCormick, Christopher N.H. Doll, and Steven B. Kraines. 2014. "Beyond the Third Mission: Exploring the Emerging University Function of Co-Creation for Sustainability." *Science and Public Policy* 41, no. 2 (April): 151–79. https://doi.org/10.1093/scipol/sct044.

UIS (UNESCO Institute for Statistics). n.d.a. "Glossary." Accessed July 6, 2020. http://uis.unesco.org/en/glossary.

UIS (UNESCO Institute for Statistics). n.d.b. "Welcome to UIS.STAT." Accessed July 6, 2020. http://data.uis.unesco.org/.

ULSF (University Leaders for a Sustainable Future). 2015. "Report and Declaration of the Presidents Conference (1990)." http://ulsf.org/report-and-declaration-of-the-presidents-conference-1990/.

ULSF (University Leaders for a Sustainable Future). 2021. "Talloires Declaration Signatories List." http://ulsf.org/96-2/.

ULSF (University Leaders for a Sustainable Future). n.d. "Talloires Declaration." Accessed May 24, 2021. http://ulsf.org/talloires-declaration/.

UNCED (United Nations Conference on Environment and Development). 1992. "Agenda 21." United Nation Conference on Environment and Development, Rio de Janeiro, Brazil,

June 3 to 14, 1992. New York: United Nations. https://sustainabledevelopment.un.org/outcomedocuments/agenda21.

UNDESA (United Nations Department of Economic and Social Affairs). n.d. "Sustainable Development Goals." Accessed July 3, 2020. https://sustainabledevelopment.un.org/topics/sustainabledevelopmentgoals.

UNESCO (United Nations Educational, Scientific and Cultural Organization). 2002. *Education for Sustainability: From Rio to Johannesburg: Lessons Learnt from a Decade of Commitment.* https://unesdoc.unesco.org/ark:/48223/pf0000127100.

UNESCO (United Nations Educational, Scientific and Cultural Organization). 2014. *Shaping the Future We Want: UN Decade of Education for Sustainable Development (2005–2014) Final Report.* Paris: UNESCO. https://sustainabledevelopment.un.org/index.php?page=view&type=400&nr=1682&menu=35.

UNESCO (United Nations Educational, Scientific and Cultural Organization). n.d. "UN Decade of ESD." Accessed July 3, 2020. https://en.unesco.org/themes/education-sustainable-development/what-is-esd/un-decade-of-esd.

Universities Canada. 2016. *2016 Directory of Canadian Universities.* 50th edition. Ottawa: Universities Canada.

Universities Canada. 2020. *2020 Directory of Canadian Universities.* 54th edition. Ottawa: Universities Canada.

University Global Coalition. 2021. "About Us." https://universityglobalcoalition.org/about/.

University of Washington. n.d. "wəɫəbʔaltxʷ—Intellectual House." Accessed July 17, 2020. https://www.washington.edu/diversity/tribal-relations/intellectual-house/.

Vaughter, Philip, Marcia McKenzie, Lauri Lidstone, and Tarah Wright. 2016. "Campus Sustainability Governance in Canada: A Content Analysis of Post-Secondary Institutions' Sustainability Policies." *International Journal of Sustainability in Higher Education* 17 (1): 16–39. https://doi.org/10.1108/IJSHE-05-2014-0075.

Vaughter, Philip, Tarah Wright, and Yuill Herbert. 2015. "50 Shades of Green: An Examination of Sustainability Policy on Canadian Campuses." *Canadian Journal of Higher Education* 45 (4): 81–100.

Vincent, Shirley, and Stephen Mulkey. 2015. "Transforming US Higher Education to Support Sustainability Science for a Resilient Future: The Influence of Institutional Administrative Organization." *Environment, Development and Sustainability* 17, no. 2 (April): 341–63. https://doi.org/10.1007/s10668-015-9623-4.

Vincent, Shirley, Stevenson Bunn, and Sarah Stevens. 2013. *Sustainability Education: Results from the 2012 Census of U.S. Four Year Colleges and Universities.* Washington, DC: National Council for Science and the Environment.

Vincent, Shirley, Sumedha Rao, Qiyuan Fu, Katt Gu, Xiao Huang, Kaitlyn Lindaman, Elishiva Mittleman, Kien Nguyen, Rachael Rosenstein, and Young-Jun Suh. 2017. *Scope of Interdisciplinary Environmental, Sustainability, and Energy Baccalaureate and Graduate Education in the United States.* Washington, DC: National Council for Science and the Environment.

Wals, Arjen E.J. 2009. *Review of Contexts and Structures for Education for Sustainable Development 2009: Learning for a Sustainable World.* Paris: UNESCO. https://unesdoc.unesco.org/ark:/48223/pf0000184944.locale=en.

Wals, Arjen E.J. 2012. *Shaping the Education of Tomorrow: 2012 Full-length Report on the UN Decade of Education for Sustainable Development.* Paris: UNESCO. https://unesdoc.unesco.org/ark:/48223/pf0000216472.locale=en.

Wals, Arjen E.J., and John Blewitt. 2010. "Third-Wave Sustainability in Higher Education: Some (Inter)National Trends and Developments." In *Sustainability Education: Perspectives and Practice across Higher Education,* edited by Paula Jones, David Selby, and Stephen Sterling, 55–74. New York: Earthscan.

WCED (World Commission on Environment and Development). 1987. *Report of the World Commission on Environment and Development: Our Common Future.* http://www.un-documents.net/our-common-future.pdf.

White, Stacey Swearingen. 2014. "Campus Sustainability Plans in the United States: Where, What, and How to Evaluate?" *International Journal of Sustainability in Higher Education* 15 (2): 228–41. https://doi.org/10.1108/IJSHE-08-2012-0075.

Whitney, Mary K. 2016. "Declarations and Commitments: The Cognitive Practice of Sustainability Agreements." In *The Contribution of Social Sciences to Sustainable Development at Universities,* edited by Walter Leal Filho and M. Zint, 89–105. Switzerland: Springer International Publishing.

Wiek, Arnim, Lauren Withycombe, and Charles L. Redman. 2011. "Key Competencies in Sustainability: A Reference Framework for Academic Program Development." *Sustainability Science* 6, no. 2 (July): 203–18. https://doi.org/10.1007/s11625-011-0132-6.

Wiek, Arnim, Angela Xiong, Katja Brundiers, and Sander van der Leeuw. 2014. "Integrating Problem- and Project-Based Learning into Sustainability Programs: A Case Study on the School of Sustainability at Arizona State University." *International Journal of Sustainability in Higher Education* 15 (4): 431–49. https://doi.org/10.1108/IJSHE-02-2013-0013.

Wright, Tarah. 2004. "The Evolution of Sustainability Declarations in Higher Education." In *Higher Education and the Challenge of Sustainability,* edited by Peter Blaze Corcoran and Arjen E.J. Wals, 7–19. Dordrecht: Kluwer Academic Publishers.

Wright, Tarah. 2013. "Stepping Up to the Challenge—The Dalhousie Experience." In *Higher Education for Sustainability: Cases, Challenges and Opportunities across the Curriculum,* edited by Lucas F. Johnson, 201–13. New York: Routledge.

Wright, Tarah, and Heather Elliott. 2012. "Canada and USA Regional Report." In *Higher Education in the World 4. Higher Education's Commitment to Sustainability: From Understanding to Action,* edited by Global University Network for Innovation (GUNi), 140–53. Basingstoke, Hampshire: Palgrave Macmillan.

Yarime, Masaru, Gregory Trencher, Takashi Mino, Roland W. Scholz, Lennart Olsson, Barry Ness, Niki Frantzeskaki, and Jan Rotmans. 2012. "Establishing Sustainability Science in Higher Education Institutions: Towards an Integration of Academic Development, Institutionalization, and Stakeholder Collaborations." *Sustainability Science* 7, supplement no. 1 (February): 101–13. https://doi.org/10.1007/s11625-012-0157-5.

I | Administrator Point of View

2
Sustainability Thinking
A View from the "Dark Side"

ROGER EPP

FIRST, A DOUBLE CONFESSION: I am not a sustainability scholar, although I have reflected on matters of place and education, in particular what it means to live in the rural prairie West with a sense of memory, care, and inheritance (Epp 2008). Instead, although not by design, I have spent a good deal of my academic working life in senior leadership roles, including those of dean and deputy provost. The last time I said yes to such an appointment, then mentioned it to a colleague, she looked at me as if there had been a death. Academic administration is commonly known as the dark side; the longer you spend there, as I know well, the harder it is to cross back. I now write at a safe distance from institutional responsibilities. But on a pro-and-con ledger, I appreciate the lateral view and the cross-disciplinary experience of the university that is inherent in such positions—that, and the opportunity to start things, to make things happen, and to speak to their purpose.

Second, a proposition: the goal of sustaining a serious sustainability agenda in higher education, not just a marketing campaign or an institutional signature on a charter, but a culture of thinking and practice so deeply embedded that it cannot be swapped out for the next "good cause," will not be realized without the skill and commitment of senior administrators. By sustainability, I mean something like Dryzek's (2005) evolving conversation of diverse voices about how to change our systems of provision and ways of life to support our renewable life support systems. In that context, senior administrators' work ought to ensure that connections get made between curriculum, research, and operations, between the university and its place, and between the various knowledges that inhabit the academy. The skill of negotiating the institutional and political thickets of the contemporary university, of finding a way, is not at all the same as compromise or cooptation. In its own way, it is critical to the future of sustainability education. The following reflections address each of those points.

| From an administrator's standpoint, a university campus is not simply a place where learning happens and degrees are earned. It is also an intense site of energy, water, food, and paper consumption, waste generation, grounds-keeping in all seasons, and building design, demolition, construction, and renovation. There are ecological responsibilities interwoven into each of those activities. The inclination to compartmentalize academic and operational realms is now risky behaviour. While a university may offer the most compelling sustainability programs and the most enlightened general education requirements, taught by leading scholars, it will only have succeeded in preparing its students to be its own best critics, sensitive to any hint of hypocrisy, if it does not walk the talk in its campus operations. If, that is, it can even recruit those students against schools that do.

The positive way to put this is that the closer integration of academic and operational realms creates meaningful on-site opportunities for learning, experimentation, modelling, and research. While the idea of the campus as living laboratory is no longer new, the benefits in terms of building a durable sustainability agenda go far beyond the entry-level goal of reducing costs. One benefit is that it challenges the traditional status hierarchies of the university, which distinguish and rank academic in relation to support staff. Closer integration means redefining who is a knowledge holder, a potential teacher, an active researcher—who, in other words, contributes directly to and shares self-consciously in the educational mission of the university. A second potential benefit is that it offsets the usual suspicions about an institution's commitment to sustainable operations by creating greater working familiarity among professors and students with what it is actually doing.[1] Cost containment does matter for public universities facing cuts to government funding, with few alternatives other than to raise tuition. But, by itself, that motivation cannot account for the enthusiasm and curiosity that sustainability practitioners bring to their work.

When I was dean and head of the University of Alberta's newly integrated undergraduate campus, Augustana, located in a small city in a rural region southeast of Edmonton, it was a matter of pride to be the first, for example, to purchase hybrid fleet vehicles, install solar panels, restrict polystyrene packing materials, and allocate land for a student-led garden. Since then, a community–university performing arts centre has been built on campus with photovoltaic cells on all sides of the fly tower and LED-only lighting throughout the building—the first such theatre on the continent. Already in my time, sustainability represented a confluence of student energy, faculty expertise, strategic planning, and hands-on staff skills. When we piloted a local food procurement program to model a more sustainable food economy, some of the inspiration might have come from the book *Animal, Vegetable, Miracle: A Year of Food Life* (Kingsolver 2007), which we put into the hands of students, but it was the cafeteria staff's knowledge and buy-in that was most critical in making it a success. Students quickly tasted the difference. More than once in that first year, cafeteria staff got a standing ovation in the dining hall. How unusual is that? By the second year, we were sourcing most of our meats and root vegetables from the region without a significant increase in costs for students. While small campuses doubtless have advantages in achieving this kind of integration,[2] the principle is the same, large or small: in building an enduring culture of sustainability, operations matter. The role of administrative leadership is, I think, to say why that is so, again and again, to soften the near-sacred distinction between "academic" and "non-academic" budgets, and to suppress the reflex always to squeeze the latter for sake of the former.

| Place also matters. Universities are always someplace, not no place, although their world-class ambitions and the convergent logic of global rankings may have helped obscure that reality. When it comes to sustainability, they cannot wish away the challenges—or responsibilities—that arise out of particular locations, ecologies, and jurisdictions; they do

make choices, however, about how they respond to them. The general point is that sustainability challenges are seldom simply generic. They require of universities an institutional self-awareness about the particularities of place.

The University of Alberta's campuses are all located above 53 degrees latitude in a cold-climate environment with dark winters. Physical geography alone presents practical sustainability challenges that are qualitatively different, say, from those in Vancouver, Tucson, or Chapel Hill. More than that, the university is the flagship public institution in a province whose low-tax resource economy has been heavily reliant for decades on oil and gas production; whose patron-client politics, high-wage workforce, and educational patterns—for example, a low post-secondary participation rate, especially among young men, and a decided preference for fields like engineering—have been shaped by that economy; and whose oilsands reserves, having been pitched as North America's best and last energy asset, have become a big, symbolic global target for climate change and anti-pipeline activism.

In this context, sustainability is necessarily a different conversation, one in which a public, comprehensive, and research-intensive university is implicated from all angles. A University of Alberta scientist first unlocked the economic potential of the oilsands almost a century ago,[3] establishing a near-continuous research connection dedicated to making the extraction process more efficient, mitigating impacts, and experimenting with techniques for land reclamation. Some of that work has been industry funded. At the same time, the university has been home to some of the most prominent critics of oilsands development, whether in ecology, health, the social sciences, or the humanities—the kind of critics, like the late David Schindler, whose public comments carry weight, make headlines, and generate heated phone calls from government ministers or donors to senior administrators.[4] The University of Alberta Press has published recent titles across this spectrum: *Unsustainable Oil: Facts, Counterfacts and Fictions* (Gordon 2015), *Upgrading Oilsands Bitumen and Heavy Oil* (Gray 2015), and *Tar Wars: Oil, Environment and Alberta's Image* (Takach 2016). Two of my colleagues in political science have published important books on the politics of oil in Alberta (Adkin 2016; Urquhart 2018); another served as director of the Alberta Climate Dialogue, a six-year experiment in citizen deliberation (see Hanson and Kahane 2018). Meanwhile, the university has educated nurses and teachers to meet labour-force needs in the oilsands regional city of Fort McMurray. Our planning professors study boom-town dynamics; our sociologists study community impacts; our forest ecologists study wildfires (before and after the devastation of spring 2016). In doing so, they move beyond caricatures to relationships and local complexities. They respond, as scholars ought, to the place in which they live.

In this context, too, the university has attempted to position itself as a key player in facilitating the province's transition to a more diverse, perhaps even a post-oil knowledge economy, and as a centre of world-class research excellence in low-carbon energy futures. But, like the sinner in Augustine's *Confessions*, neither the university nor even its professors can afford to be made chaste just yet. From an administrator's standpoint, the structural weakness of the province's public finances—so dependent on a volatile and diminishing revenue stream—means that our operating and capital budgets expand and contract with the resource economy. So does the ability of students or their parents to afford tuition. In the past, we have been secretly relieved when prices rose and projects resumed. In the current context, we brace for the budgetary reckoning that will accompany a long-term decline of the industry, and the response of a government that has been unwilling to contemplate a post-oil future and all too willing to mark higher education as the safest sector, politically and ideologically, on which to impose punitive cuts (Cryderman 2021). Which is to say: sometimes the constraints and the institutional costs can loom very large.

One more thing about the University of Alberta: by history and location, it is seriously north facing—that is, it is home to researchers who work across the circumpolar region. As scientists, they have contributed to raising the international scientific alarm about melting sea ice, glaciers, and permafrost. As anthropologists, biologists, public health specialists, and engineers, they have worked alongside northern governments and Indigenous communities as they have struggled to understand and adapt to the effects of rapid climate change on land, water, vegetation, traditional diets, housing, and transportation. The further north you go, in other words, the further you get from the state of denial. Collectively, the university knows enough and carries enough of a sense of urgency to not become complacent about the ecological crisis to which sustainability education must point.

| While everyone is now in favour of sustainability, what it means and what it requires are contentious questions. They can divide campuses no less than electorates, as the University of Alberta was reminded when the announcement of an honorary degree for a prominent Canadian environmentalist during a downturn in the energy economy ignited a firestorm of public protest, even from deans. The politics of sustainability invite greener-than-thou suspicions about the university, or about the work of colleagues who serve on government-appointed environmental monitoring panels or imagine technological fixes for sustainability challenges when the real problems, others will argue, are inherent in the deeper logic of capitalism. They make boards of governors and finance vice-presidents nervous at the prospect of campus campaigns for fossil-fuel divestment.[5] All together, they reveal something of the knowledge solitudes of the contemporary university.

In a perceptive essay, sociologist Michael Burawoy poses the question of how to redefine the public university in its own defence, against the pressures of regulation and commodification—in other words, performance indicators, global rankings, and commercialization mandates. Those pressures, he writes, threaten to turn higher education into a mere "means for someone else's end" and to sever "its ties to national and local issues" (Burawoy 2011, 27, 35). His defence affirms the integrity of four kinds of knowledge on which the university's work rests. The first is professional or specialist knowledge, which is judged by the peer-review standards of the academy. The second is the instrumental or problem-solving application of this knowledge outside the academy as policy knowledge in the external worlds of industry or government. The third, critical knowledge, stands in tension with the second. Transcending specializations, it represents the "conscience of the university." It asks reflexively: What is the university for? What are its core values? Knowledge for whom and for what purpose? In these questions, critical knowledge is informed by and contributes, in the form of advocacy, to public knowledge: what serves a greater good, and what threatens it. Burawoy's (2011) university holds those four kinds of knowledge in balance within and across all of its disciplines, in research as well as in teaching. It requires them to acknowledge and live out their interdependence, to uphold the university's "precarious autonomy," its coherence as a *uni*versity, and its ability to serve as a critical public sphere (Burawoy 2011, 40). In isolation, these knowledges are subject to capture or to the irrelevance of an echo chamber.

I have outlined Burawoy's argument at length because it maps very well onto the sustainability question in higher education, which is, at once, instrumental and critical reflexive. Although he does not go this far, his argument also leaves a role for academic leadership—deans, provosts, presidents—in balancing different knowledges and in convening conversations across their solitudes, sometimes leading and provoking them. This is not an exclusive role. But those with the authority to speak for the university make choices constantly about what to communicate, how, and when to risk leadership capital to serve as a champion. Those with the authority to spend its discretionary funds make choices about the cross-faculty initiatives and incentives, workshops, and grants they will support in shaping an inclusive, informed, and reflexive conversation. Those choices are important to the future of sustainability in higher education.

| In closing, I want to propose a fifth kind of knowledge that is essential in building a deep, dynamic culture of sustainability thinking and practice within universities. Aristotle (1980, bk. 6) once called it "phronesis." By that he meant a practical wisdom shaped by experience and marked by a capacity for judgment, the virtue of knowing how to apply sound principles in complex situations toward some good end. Phronesis is a political rather than a theoretical wisdom. It sees a way through situations. It acts. It does not lose sight of the good, which it can also apprehend and articulate. It understands the importance of timing. It understands decision-making processes and structures. In matters of sustainability, it might resemble an incremental radicalism—pragmatic, opportunistic, but also relentlessly unfinished.

My point is neither to equate phronesis automatically (and only) with university administration nor to invoke it as an insulating shield, an excuse, the kind that complains: "You don't understand the pressures I'm under, the heat I take from government, from donors; you need to be in my shoes before you criticize." I know academic leaders who have acted courageously and at some cost to keep sustainability issues in front of their campuses. Indeed, there can be costs, especially in an energy-dominated political jurisdiction, where oil is a matter of identity as much as a commodity or revenue stream, and where the price of "disloyalty" can be high for public universities. But the complex institutional and societal circumstances in which all universities operate, as described above, require that phronesis be part of the broad campus conversation convened around sustainability; for it is not so great a stretch to read Aristotle to say that practical wisdom can be a quality of deliberative bodies, of citizens, as well as of those chosen to lead.

To build an institutional culture of sustainability thinking and practice, campus operations matter. Place matters; it must be owned in all its opportunities and limits. Conversation across knowledge solitudes matters; it must be convened, modelled, and encouraged. In every respect, articulate, principled leadership matters. In the times in which we live, it is no longer optional.

Notes

1. At the University of Alberta, the Office of Energy Management was created four decades ago in the aftermath of the first OPEC oil-price shocks. See the University of Alberta's Vice President Facilities and Operations website at https://www.ualberta.ca/vice-president-facilities-operations/projects/energy-management-and-sustainable-operations/index.html; and the University of Alberta Sustainability Council's website at https://www.ualberta.ca/sustainability/index.html.
2. Within the Council of Public Liberal Arts Colleges, which the Augustana Campus joined as the first Canadian member, the University of Minnesota at Morris has set the operations standard with its commitments to wind energy, biomass energy, and local foods—toward the goal of carbon neutrality. Those commitments have enjoyed student support and administrative leadership. See the university's "Sustainability Goals and Initiatives" at https://morris.umn.edu/sustainability/sustainability-goals-initiatives.
3. A brief biography of Karl Clark can be found at http://www.history.alberta.ca/energyheritage/sands/unlocking-the-potential/the-scientific-and-industrial-research-council-of-alberta/karl-clark.aspx. The Institute for Oil Sands Innovation, which has received significant financial support from companies like Imperial Oil, is the most visible representation of this linkage at the University of Alberta; see https://iosi-alberta.ca/. For a very different, critical kind of energy research also based at the university, but in the arts, see the Petrocultures Research Group website at http://petrocultures.com/.

4. I am well aware of the political-linguistic contention between "oilsands" and "tar sands" and the ways in which those words, especially the latter, are used to signal positions. My preference for the former in this chapter reflects an agnostic's view that labels are not what's fundamentally at stake in the debate and that they can, if anything, make the public conversations that need to happen even more difficult.

5. Tellingly, in fall 2016, the Canadian Association of University Business Officers held a stand-alone, two-day workshop: "Building a Toolkit for Effective, Ethical and Responsible Responses to Divestment Campaigns." Since then, divestment from the fossil fuel sector has been the focus of high-profile campus campaigns. For example, at McGill, the senate passed a motion in 2018 in favour of full divestment, which a board committee rejected in a lengthy report that recommended a gradualist approach, although still in the direction of "a more sustainable and less carbon-intensive investment strategy" with specific "decarbonisation targets" (McGill 2019, n.p.). At the University of British Columbia, the board endorsed its president's declaration of a climate emergency in December 2019 and subsequently committed to "full divestment as soon as possible" on the understanding that "the continued operation of the fossil fuel industry is discordant with a climate safe future" (University of British Columbia 2020, n.p.). Such intense divestment debates suggest a number of things. First, place clearly matters; similar campaigns would be a non-starter with public university boards in Alberta. Second, the focus on university endowments is not going away. Third, although the language of sustainability is elastic, it is increasingly a frame for gradualist positions over and against the more urgent language of climate emergency.

References

Adkin, Laurie E., ed. 2016. *First World Petro-Politics: The Political Ecology and Governance of Alberta*. Toronto: University of Toronto Press.

Aristotle. 1980. *The Nicomachean Ethics*. Translated by David Ross. Oxford: Oxford University Press.

Burawoy, Michael. 2011. "Redefining the Public University: Global and National Contexts." In *A Manifesto for the Public University*, edited by John Holmwood, 27–41. London: Bloomsbury Academic.

Cryderman, Kelly. 2021. "Alberta Cuts Postsecondary Funding in New Provincial Budget." *The Globe and Mail*, February 27, 2021. https://www.theglobeandmail.com/canada/alberta/article-alberta-cuts-postsecondary-funding-in-new-provincial-budget/.

Dryzek, John S. 2005. *The Politics of the Earth: Environmental Discourses*. 2nd ed. New York: Oxford University Press.

Epp, Roger. 2008. *We Are All Treaty People: Prairie Essays*. Edmonton: University of Alberta Press.

Gordon, Jon. 2015. *Unsustainable Oil: Facts, Counterfacts and Fictions*. Edmonton: University of Alberta Press.

Gray, Murray R. 2015. *Upgrading Oilsands Bitumen and Heavy Oil*. Edmonton: University of Alberta Press.

Hanson, Lorelei L., and David Kahane. 2018. "Introduction: Advancing Public Deliberation on Climate Change and Other Wicked Problems." In *Public Deliberation on Climate Change: Lessons from the Alberta Climate Dialogue*, edited by Lorelei L. Hanson, 3–31. Edmonton: AU Press.

Kingsolver, Barbara. 2007. *Animal, Vegetable, Mineral: A Year of Food Life*. With Steven L. Hopp and Camille Kingsolver. New York: HarperCollins Publishers.

McGill University. 2019. "McGill Board of Governors Receives Recommendations to Decrease Carbon Footprint of Investment Portfolio." *McGill Newsroom*, December 3, 2019. https://mcgill.ca/newsroom/channels/news/mcgill-board-governors-receives-recommendations-decrease-carbon-footprint-investment-portfolio-303083.

Takach, Geo. 2016. *Tar Wars: Oil, Environment and Alberta's Image*. Edmonton: University of Alberta Press.

University of British Columbia. 2020. "Update: Next Steps Following Climate Emergency Declaration and Commitment to Divestment." January 10, 2020. https://vpfo.ubc.ca/2020/01/ubc-update-moving-toward-divestment/.

Urquhart, Ian. 2018. *Costly Fix: Power, Politics, and Nature in the Tar Sands*. Toronto: University of Toronto Press.

3

Sustainability Scholarship and Education
Opportunities and Strategies for Success

CHRISTOPHER G. BOONE

SUSTAINABILITY AS AN ACADEMIC FIELD OF INQUIRY has moved from curious side projects to become a more central concern of colleges and universities. At my own institution, sustainability is a principle that pervades the entire university, guiding everything from research and education to practice and community engagement. For the purposes of this chapter, I define sustainability as improving human well-being and ensuring social equity for present and future generations while safeguarding the planet's life-supporting ecosystems.

The ideas in this chapter reflect my seven years of experience as dean of the School of Sustainability at Arizona State University (ASU). In this role, I have collaborated with other deans, faculty, staff, students, and external partners to continuously improve the nature of sustainability education, both formal and informal. As a board member for the Global Council on Science and the Environment and a committee member for a National Academies of Science, Engineering, and Medicine initiative on sustainability in higher education (Kapuscinski et al. 2020), I have worked with colleagues to create a consensus on key competencies and capacities in sustainability for colleges and universities. My research has always been intentionally interdisciplinary and collaborative. For more than two decades, I have used environmental justice as a helpful research framework. I was drawn to sustainability because of its systems and integrative nature, its normative focus that includes justice as a core principle, and its courage to focus on intervention and solutions to disrupt the status quo. From my undergraduate studies at Queen's University down to the present, I have directed most of my research time toward understanding the uneven consequences of urbanization while also celebrating the energy and diversity of urban life. It is my firm belief that although urbanization has created many of the stresses on human and planetary systems, it will be sustainable urbanization that will lead us to a sustainable future.

When Arizona State University formerly established the School of Sustainability in 2006, it was a rather lonely club, but sustainability programs grew quickly soon after. As of spring 2020, the Association for the Advancement of Sustainability in Higher Education (AASHE) had listed more than 2,800 sustainability-related programs submitted by colleges and universities from twenty-seven countries, all states and territories in the US, and nine Canadian provinces. Sustainability as an academic enterprise has reached critical mass and will continue to be adopted as a formal field of study and research. However, as a relatively young field, it should continue to critically reflect on what makes it valuable or different from existing programs. Sustainability scholarship and education must evolve to meet changing needs for the university and external communities, or risk becoming irrelevant. Below, I outline some promising opportunities for sustainability scholarship and education. I discuss strategies to achieve success and intersperse examples from the School of Sustainability at Arizona State where possible to illustrate the application of the themes I cover in this chapter.

Future Opportunities for Sustainability Scholarship

Sustainability is an outcome-driven science with an explicit attention to interventions that lead to desirable futures. Normative principles related to equity, justice, values, cultures, wants, and needs, among others, motivate the interventions and become part of the scientific framing of sustainability scholarship (Horcea-Milcu et al. 2019; Klinsky et al. 2017). This blending of desired outcomes and science can be unsettling to some scholars and liberating to others. Regardless of their proclivities regarding the proper role for science and universities, sustainability scholars are motivated by the urgent need for change and the importance of interventions that lead to a better future (Miller et al. 2014). Similar to medical and health professions, sustainability scholars embrace the notion of science-based interventions to improve conditions for people, as well as the planet (Abson et al. 2017). Sustainability analogies to health sciences are helpful, yet limited, since good health is somewhat easier to define than a desirable future for the complex social–ecological–technical system that is the sustainability scholar's patient. However, the health sciences and sustainability share the idea that intervening rather than only understanding the nature of the problem is a fundamental core mission. Sustainability and medicine are goal-driven sciences and practices that are committed to improving the lot of humanity, now and into the future.

One of the most difficult challenges for sustainability is defining a desirable future. How to incorporate the multitude of visions that depend on the values and ideals of individuals, communities, societies, and states, and the inherent conflicts of interest between them is a formidable exercise. A related concern is setting a course for a desirable future that has unintended consequences, making things worse rather than better. An example is the drive for biofuel production in the 1990s, undertaken with the intent to create carbon-neutral energy, but a strategy, now entrenched, that has not lived up to its promises (Searchinger and Heimlich 2015). Interventions, even well intentioned, can create path dependencies that set systems on courses that are hard and expensive to correct. More work needs to be done on developing interventions that are agile, responsive, and adaptive to changing conditions, knowledge, or ideas (Tellman et al. 2018).

Given the challenge of defining the future we want, how can sustainability scholarship proceed? One strategy is to begin with frameworks or principles that have reasonably wide acceptance. At Arizona State University, we have surveyed our designated 550 sustainability scientists and scholars to gauge their research alignment and interests with the seventeen United Nations (UN) Sustainable Development Goals (SDGs).[1] Several reasons motivated this exercise. First is the scale and magnitude of the stakeholders that these goals represent as they were unanimously approved by all member countries of the UN in January 2016. This is the closest the world has come to adopting a global consensus for a future vision. Second is the relative success of the Millennium Development Goals and the Millennium Ecosystem Assessment in setting targets, motivating constituencies, and making real progress toward them. This brings a certain degree of faith that sustainability agendas can make a positive difference, and many regions are demonstrating traction toward the goals (Xu et al. 2020; Ritchie, Roser, Ortiz-Ospina, and Mispy 2018). Third is the global scope of the SDGs. Sustainability is ultimately a global scale concern. Strategies employed at the local level, even with the best motivations and care, can potentially undermine the ability of other locales to meet their own sustainability goals or in aggregate could make conditions worse for all, feeding back to undermine many or all local efforts (Seto et al. 2012). Fourth is the ability of the SDGs to organize research and teaching interests across a broad spectrum of disciplines and expertise at the university. The survey results allow us to identify clusters of experts who are working on specific goals (e.g., responsible consumption and production, gender equality), and to find ways for them to work together on achieving the goals and the subset targets.

Integrated Systems Approach

A strength of sustainability is the focus on integrated systems. A systems approach improves understanding of the interacting dynamics of individual components and is a smart way to plan interventions. In the United States, the National Science Foundation launched several calls for proposals that support integrated systems research, such as the Dynamics of Coupled Natural and Human Systems program, Science, Engineering, and Education for Sustainability, Sustainability Research Networks, Dynamics of Integrated Socio-Environmental Systems, and Innovations at the Nexus of Food, Water and Energy Systems.[2] In the US southwest region where I live, it is difficult—and foolhardy—to separate food from water from energy. In the simplest understanding, food is embedded with water, which is embedded with energy. Electricity and pumps lift water from the Colorado River to run by gravity feed to the Metro Phoenix region and then on to Tucson and points further south. The energy embedded water is used to grow food and feed (including hay that is transported to horse stables and for cattle feed in Saudi Arabia; see NPR 2015), which requires energy and water to transport, store, process, deliver, and dispose of them. Any sustainability strategy to reduce energy consumption should therefore take into account the role and implications of changes to the food–water–energy nexus. This includes how to set the boundaries of interventions, since food grown in Arizona that is embedded with water and energy may be consumed elsewhere in the country or other parts of the globe. California, which is prone to drought, exports about 600 trillion gallons of "virtual" water (500 gallons per resident per day) that is embedded in and required to grow agricultural commodities shipped out of state (Davidow and Malone 2015). A great deal of work needs to be done on full cost accounting to understand the stocks and flows of environmental and social benefits and burdens.

The Anthropocene is largely a product of the evolution of human beings into an urban species. Cities currently account for 75 percent of greenhouse gas emissions, 65 percent of energy consumption, and 75 percent of global economic activity as a contribution to GDP (Mi et al. 2019; Solecki et al. 2015; Acuto, Parnell, and Seto 2018). Despite the proliferation of slums, moving to cities is one of the surest ways for families to escape poverty and to access services such as education and health care. Urbanization, barring catastrophe, will continue apace and cities will house another 2.5 billion people, or two-thirds of the world's population, by mid-century. Africa and Asia will urbanize faster than any other region and account for 90 percent of the world's urban growth. India will add nearly 420 million new urban residents by 2050, the highest projected growth of any country, followed by China with 255 million and Nigeria with 189 million (UN DESA 2019). Urbanization is the dominant human force on the planet, with the demand and activities of cities and urban dwellers driving everything from natural resource extraction and land use change to biogeochemical cycling and ocean acidification. A key area of research and practice is finding ways to direct urbanization to achieve sustainability goals for urban dwellers and the planet as a whole (Boone et al. 2014). Even with shocks such as the coronavirus pandemic that began in 2020, which has had very high incidence rates in densely populated cities, it is highly unlikely that urbanization will go in reverse. Using the energy and investment in urbanization to meet global sustainability goals is an area of huge opportunity for sustainability scholars and their work with partners and practitioners (Acuto, Parnell, and Seto 2018).

Any efforts in sustainability must tackle the issue of how to govern the commons. This is particularly acute for resources that are shared globally, including the oceans and the atmosphere, where actors working in their self-interest undermine prosperity and well-being for all. Exactly what mechanisms or institutions will be required to govern the health of the oceans or atmosphere remains an open question, but it is a fundamental area in the sustainability project. The latest Conference of Parties for the UN Framework Convention on Climate Change pointed to the importance of setting goals from a bottom-up approach as an alternative to top-down regulation (Ostrom, Janssen, and Anderies 2007). The sticky question, raised in the climate justice literature, is how compensation for past greenhouse gas emissions will factor into the implementation of country-specific goals, termed Intended Nationally Determined Contributions (Klinsky and Winkler 2014; Klinsky et al. 2017). Although everyone faces dire consequences from climate change, the questions of equity cannot be eliminated from the conversation or from strategy. This is a fertile area of inquiry from sustainability scholars who can employ their expertise in law, political science, justice studies, and environmental economics.

The arts, humanities, and design are often an afterthought in sustainability scholarship, brought in to help communicate or disseminate the findings of scientists. This is a mistake and a wasted opportunity. Successful pathways to a sustainability transition ultimately depend on understanding the human condition and finding ways to change or motivate human behaviour and decisions (Adamson and Davis 2016). My geology friends, with the benefit of a very long-time perspective, note that the planet will go on, with or without us, as it has before, and will for another few billion years (Weisman 2007). Colleagues from the humanities, arts, and design should be at the table at the beginning, helping to formulate the research designs so that collectively we are asking critical questions. The desired, sustainable future is at its heart a design question. It is not simply the monitoring and evaluation of current systems. How might science experiments combine with the tools of design to develop, for example, a prototype for the next 420 million urban residents in India or for retrofitting aging cities in North America, Europe, and Latin America to meet the UN SDGs? How can art encapsulate the complexity of the future in ways that are comprehensible to all people and provide roadmaps for how to get there? What can we learn from history about past mistakes and successes that can inform future decisions? How will literature inform the sustainability narrative and inspire change? Will climate fiction change more minds and alter more actions than the efforts of the United Nations Framework Convention on Climate Change?

Opportunities for Sustainability Education

A core mission of the university is to educate and foster learning. It is the biggest lever in the toolkit of the university, where concerted and quality efforts can have near to immediate impact. Research and scholarship play critical roles, too, but often the effects on sustainability outcomes can be diffuse, difficult to track, and delayed. At my own institution, the greatest discernable impacts we are having on sustainability are on the education of students and the great work they are doing as alumni. One of our master's degree graduates started his own company, Carbon Roots International, in Haiti that turns crop residue into biochar. The biochar is used as a soil amendment and also to produce charcoal at a lower cost and with the same or better performance than traditional wood charcoal. The result is improved soil structure and quality and reduced stress on an already deforested landscape. The company employs local workers to produce the biochar, and women and children act as resellers for income. It is an excellent example of how a former student educated in sustainability tackled a serious problem with a solution that benefits the very poor and the stressed ecological services upon which they rely.

Sustainability degrees at ASU use a competency framework to guide the development of courses, curricula, programs, and assessments. Competencies are defined as "a functionally linked complex of knowledge, skills, and attitudes that enable successful task performance and problem solving" (Wiek, Withycombe, and Redman 2011, 204). For sustainability, we identified five core sustainability competencies—systems, anticipatory, normative, strategic, and collaborative thinking. Systems thinking is the ability to understand the dynamics of integrated, complex socio-environmental-technical systems. This is fundamental for understanding how the world works and for avoiding the traps and consequences of simple linear thinking. Anticipatory thinking is the ability to envision plausible and desirable future states (McPhearson, Iwaniec, and Bai 2016). Normative thinking encourages our students to think about what "should be" rather than only what "could be" included in a vision for a future we want. It also teaches students the importance of values for motivating people and communities, including ethics and justice for enduring solutions and doing the right thing, and for recognizing how our own values shape our visions for the future. The strategic thinking competency is the ability to craft, assess, and implement strategies to get us to a desired future. The collaborative competency is the ability to effectively engage stakeholders, develop empathy, communicate effectively, arbitrate, mediate, and use other approaches for sustainability problem solving and solutions generation. This competency framework has proven to be a robust and relevant way to educate students in sustainability at all levels. A report by the National Academies of Sciences, Engineering, and Medicine has recommended a competencies-based approach as one means for strengthening sustainability education at higher education institutions (Kapuscinski et al. 2020).

At ASU, we used these competencies to develop program level learning outcomes (PLLO), and all courses map to these outcomes at designated low, medium, and high levels. This exercise allowed us to map out the coverage of our PLLO in order to identify gaps, which guides teaching and research investments. In turn, we created program level assessment tools using the program level outcomes. Students create digital portfolios, beginning in their first year, to self-assess their mastery of the core competencies and the PLLO. The digital portfolio is a tool that students can use to show employers the skills and content they have mastered, which is a marked improvement over a transcript or simple resume.

The coronavirus pandemic that began in 2020 forced many colleges and universities to quickly adapt remote digital learning. While the rapid transition from in-class to online learning was certainly disruptive, the crisis demonstrated some benefits of digitally enhanced education (Noetel et al. 2020). In the School of Sustainability, class attendance rose when we shifted to synchronous online courses delivery, compared to when those courses were offered in person. We also found that normally quiet students were more likely to participate in the videoconferencing format compared to the traditional classroom setting. Since lectures were recorded, this meant students could go back and review the lecture material if they misunderstood content and to study for tests and exams.

Online learning will continue to grow in all fields, with specific benefits to sustainability. The first and most critical is that online education opens the door to millions of people who might not otherwise have an opportunity to access higher education (Crow and Dabars 2020). Sustainability depends on extensive stakeholder engagement, and if students interact only with those who have the resources and time to attend on-campus courses, their worldviews will be limited. Online technologies also offer opportunities to develop distributed seminars with colleges and universities around the world. They give students a chance to speak directly with authors and practitioners, and work on joint projects with students working in different political, economic, cultural, and environmental contexts. Ideally, these initial interactions can provide the groundwork for rich and engaging global immersive experiences, such as study abroad programs and exchanges for students and faculty.

Informal Education

I see tremendous opportunities for informal education in sustainability, and at my own institution we have made substantial investments in it. Increasingly, organizations are looking to universities to provide training in sustainability, in part because many of their employees did not have the benefit of an education in sustainability, given its relatively recent adoption in higher education. Training can range in format from short, online sessions to multi-day workshops. At ASU, we see two immediate benefits. The first is that it provides revenue, which can be used to support faculty, staff, and students. The second is that it is a meaningful way to engage with institutions to assess their sustainability needs. We use these trainings and workshops to gauge the sustainability priorities of institutions and then feed that back into our own curricula. This ensures that students that we graduate are aware and ready to tackle the sustainability questions of their potential employers. A third benefit is that it builds relationships with influential organizations—those that have the capacity and ability to implement sustainability strategies in ways different from universities. For example, we have offered several workshops with the World Bank and the International Finance Corporation on climate change and investment risk. If these workshops result in a change of thinking that may influence change of action, the on-the-ground effects can be substantial. Our own faculty, staff, and students also learn how sustainability is practiced on a day-to-day basis in contexts and institutions that may not have the luxury of long-term reflection. This provides them with a chance to think about how to truly move knowledge to action based on real constraints and opportunities.

A final benefit of informal education is the ability to reach very large audiences. It is not a substitute for the deep engagement students receive in various degree programs, but it does allow universities to touch thousands or even millions who may not have the time, resources, or inclination to pursue formal education in sustainability. Urgent sustainability challenges require solutions that match the scale of the problems, and informal training could help in that regard.

K to Grey

The primary focus of universities is on the education of adults, but they can play an important role in educating all ages, from "K(indergarten) to Grey." In sustainability, this is particularly important because of the field's relative youth. Introducing sustainability thinking into elementary and secondary education prepares students for degree programs in sustainability or for a different way of framing whatever they choose to learn. Lifelong learning efforts are critical in the information age, and sustainability provides a fresh perspective on critical problems and creative solutions. In the School of Sustainability, we have welcomed opportunities to educate from K to Grey through the informal education efforts described above, online degree programs, and working with teachers. Using generous funding from the Walton Family Foundation and Wells Fargo, we created a Sustainability Teachers' Academy.[3] The purpose is not to impose yet another teaching requirement but to demonstrate the way teachers can weave sustainability principles throughout curricular and non-curricular activities. The program is a teach-the-teacher model, with the expectation that participating teachers will return to their home districts and work with teachers and students to adopt sustainability learning and practices.

Place-Based Education

Sense of place, a feeling of connectedness forged through a deep understanding of an area's cultural and natural history and characteristics, is a key sustainability principle and a potential motivator for sustainable behaviours (Grenni, Soini, and Horlings 2020; Newman and Jennings 2008; Rogers and Bragg 2012). We use place-based education approaches to help students identify their own sense of place and see how it may motivate others to look after what they care about. Similar to many other universities, ASU uses its campus as a living laboratory, coordinating sustainability projects with University Sustainability Practices, which is responsible for tracking and implementing the university's sustainability goals.[4] Faculty employ place-based education as a way of demonstrating complexity in tangible environments, providing a manageable field of view for students to explore the interactions between social, ecological, and technical systems and to seek solutions. One case I find particularly powerful is the Tres Rios artificial wetland, a facility created to treat wastewater from the Greater Phoenix region using ecosystem services rather than relying solely on hard infrastructure. Several students conducted research at this site and worked with a filmmaker and professor of practice in our school to create a documentary about this special place.[5] The opportunity to use documentary filmmaking as a way to capture Tres Rios' sense of place and explain complex sustainability issues was very enlightening for the students. A second project weaves traditional knowledge and science perspectives about Kahaluʻu Ma Kai, a sacred place on the Big Island of Hawaiʻi.[6] This virtual field trip uses 360-degree immersive technology and annotates the place with instructive videos, text, and interactive questions from cultural resource specialists at Kamehameha Schools, who partnered with ASU faculty on the project. In addition to instilling sense of place, the virtual field trip allows students in Hawaiʻi and from around the world to "travel" to sites that would otherwise be difficult to access. The virtual field trips have an explicit mission of teaching students about sustainability by blending Native Hawaiʻian cultural-based and STEM-based approaches. Whether or not virtual field trips can create something close to what a "real" field trip creates regarding sense of place is an open question. However, it is one way to capture essences of places that are under threat, sensitive to disturbance, or difficult to access. It might also be a means of creating a more global sense of place, giving children and adults an opportunity to learn something meaningful about places halfway around the world.

Strategies and Tactics

Avoid the Temptation to Own Sustainability in a Single College or School
At my institution, sustainability is not "owned" by the School of Sustainability. I believe it would be a grave mistake to view it this way. To understand the complexity of sustainability challenges and build creative solutions requires the breadth of expertise that universities, along with their partners, have to offer. First and foremost, this means that success in sustainability scholarship and education requires a willingness and commitment to work beyond the traditional boundaries of universities defined by fields of study, majors, degree programs, departments, schools, colleges, and reporting structures. Perfect fluidity may not be possible, but boundary spanning (and breaking) should be a guiding goal. Universities are not entirely unfamiliar with these concerns. Debates and struggles over who owns the environment as a field of study have been raging since the 1970s. More recently, entrepreneurship has entered the realm of contested territory. Others argue about who has the exclusive authority to teach mathematics or writing. While there can be clear budgetary incentives that shape these arguments about territory and ownership, most faculty—and certainly the students they serve—recognize that these cross-cutting themes, principles, and ways of knowing and expression are strengthened through interdisciplinary education and application. In the School of Sustainability, we do whatever we can to support sustainability programs across campus by offering seats in our courses to other majors, cross-listing our courses with those of other units, exchanging faculty to teach in various programs, offering our courses to aid in the development of sustainability specialties in other degrees, or simply providing verbal and written support for sustainability initiatives in other schools and programs. It helps that student credit hours (SCH), which drive a good deal of budgets at ASU, follow whoever pays the instructor's salary. This means that a faculty member from the School of Business can teach a critical course to our students in the School of Sustainability and the SCH

will flow back to the dean of the School of Business. This provides greater freedom for finding expertise anywhere on campus rather than having to contain it within one college or school. It also reinforces the notion that sustainability is not owned by a single college or school.

Modify Promotion and Tenure Guidelines
Academic traditions can be difficult to change, but perhaps none so much as faculty promotion and tenure. Colleges and universities hire tenure-track faculty with the expectation of their excellence in research, teaching, and service. Promotion and tenure committees typically separate their assessments into discrete categories and expect excellence—depending on the nature of the institution—in research, teaching, or both, with service being a distant third (especially for junior faculty). External letter writers are usually asked to judge the candidate's research record and sometimes their service, but rarely their teaching. All of this leads to an artificial fracturing of a faculty member as a whole person. This may be particularly problematic in sustainability, which depends on integrated systems approaches to identify and solve problems with an identified set of stakeholders. Even when using classic research skills, a sustainability scientist or scholar may be motivated by a very applied problem and deliver the solution as a report or training exercise that does not fall under the traditional peer review definition of research. Clearly, the peer review process is valuable and has advanced knowledge profoundly over the last century. At the same time, the sense of urgency around sustainability issues means that faculty are motivated to use their research and teaching talents to search for solutions that are practical and understandable, and that will have near-term positive effects.

Engaging more faculty in sustainability requires some shifts in university incentives and structures. One mechanism is changing the promotion and tenure guidelines. In the promotion and tenure guidelines of the School of Sustainability, we recognize reports and so-called "grey" literature as an important form of scholarship, albeit not a replacement for peer-reviewed scholarship. We also give junior faculty the choice of demonstrating excellence using disciplinary standards (e.g., economics, biology, history, engineering) or by making a broader contribution to the field of sustainability as an outcome-oriented science. This by no means dismantles the traditions of promotion and tenure (some junior faculty are understandably risk averse, especially if they want to move on to a more narrow, disciplinary academic position elsewhere), but it does start to move in the direction of changing what is valued as scholarship, teaching, and service.

Let Students Be
Sometimes the best advice for promoting sustainability programs at universities is to simply get out of the way of students. While turf battles rage on among faculty, staff, and administrators, students on university campuses hunger for an education and set of experiences that will help them understand the complex world they are inheriting and what they can do to fix it and make it a better place. A fifty- or sixty-year time horizon is more than an abstraction for young people—it represents the decades they expect to live. They have a vested interest in making sure they can enjoy long, meaningful lives supported by healthy and thriving social and ecological systems. Supporting student organizations that focus on sustainability is a good way to kick-start more formal engagements in college curricula and research programs. Using the campus as a living laboratory creates a somewhat safe-to-fail environment for students and others to experiment with sustainability problem solving and solutions generation. Providing students with sustainability-related internships is a way to bring back real-world concerns to university sustainability programs. Offering introductory courses, certificate programs, or minors in sustainability allows students to vote with their feet and demonstrate to faculty and administrators the demand and desire for training and education in this field.

Focus on Inclusive Well-Being
In a primer on sustainability, Matson, Clark, and Andersson (2016, 15) define sustainability as the pursuit of "inclusive social well-being." It is social in the sense that sustainability is about more than individual needs and desires. Inclusive refers to equity or fairness within and between generations. Well-being means more than good health—it refers to a bundle of characteristics and conditions, material and otherwise, that lead to a good quality of life. Implied but not stated is that inclusive social well-being depends on the health and proper functioning

of earth's life-supporting systems. This definition is very much in the "big tent" spirit of sustainability, inviting the participation of a broad set of constituents. If done well, a broad sustainability definition can create an all-in approach to sustainability that engages faculty, staff, and students from across the university and encourages external partners to engage as well. By inviting many to participate, an inclusive, all-in strategy can build a critical mass of ownership that may be necessary for launching and maintaining new programs.

Find and Cultivate Good Leaders
Strong leaders—especially at the provost or presidential level—who openly and publicly support sustainability initiatives can greatly improve chances for success, particularly if programs use the big tent approach described above. Because sustainability is difficult for any single college, school, or department to "own," a leader who has responsibilities at an institutional level is very important for advancing successful sustainability programs. Finding or cultivating leaders with a serious commitment to interdisciplinary, outcome-oriented research, teaching, and outreach is an important recommendation for colleges and universities about to set off on a sustainability pathway (Gordon et al. 2019; Boone et al. 2020).

Teach Hope, Agency, and Responsibility
Hope, agency, and responsibility are three principles for any sustainability program (Boone 2015). One of the reasons why students are attracted to sustainability as a field is because it recognizes that the future is not fixed. Whereas many environmental science programs focus solely on the nature of problems, often casting a bleak future for people and the planet, sustainability rests on the notion that there is an opportunity to build a better future. This is not to downplay the seriousness and urgency of sustainability challenges such as climate change, poverty, or biodiversity loss, but without hope, there is little to motivate people to move beyond the status quo. The second principle of agency is equally critical. It suggests that individuals and groups have the power to effect positive change. Taking responsibility for our actions and using the privilege of responsibility to work for positive change is a third motivator for students. Waiting for others to intervene will likely perpetuate

the status quo or at least slow down the transition necessary for building a sustainable future. When individuals, especially those who have the privilege to receive a university education, take responsibility and combine it with hope and agency, progress toward sustainability is a very real prospect.

Conclusion

The roadmap to building sustainability programs across North America will accommodate multiple pathways, as it should. Colleges and universities will start at different origins and will have to navigate using various assets while overcoming distinctive roadblocks. Some programs will start small, as grassroots experiments, while others will begin as university-wide, all-in efforts supported by strong leadership. All programs share a similar goal of high-quality sustainability education, research, and outreach to create a better future. It is critical that we build supportive networks of sustainability programs, rather than cut-throat competition, to ensure that we all succeed in reaching our destination.

Notes

1. See the UN SDGs at https://sustainability.asu.edu/sustainable-development-goals/.
2. Each of these calls can be found on the National Science Foundation website at https://www.nsf.gov/.
3. See more information on the Rob and Melani Walton Sustainability Teachers' Academies program at ASU at https://sustainability.asu.edu/sustainabilitysolutions/programs/teachersacademy/.
4. See more on ASU's sustainability and operations practices at https://sustainability.asu.edu/operations/.
5. To see *Tres Rio Wetland*, created by Suzanne Jumper, Yeowon Kim, and Johnathan Rugg, go to https://vimeo.com/147837180.
6. See "Hawai'i Kahalu'u Ma Kai" at https://aelp.smartsparrow.com/v/open/mtbw5q95.

References

AASHE (Association for the Advancement of Sustainability in Higher Education). 2020. "Academic Programs." https://hub.aashe.org/browse/types/academicprogram/.

Abson, David J., Joern Fischer, Julia Leventon, Jens Newig, Thomas Schomerus, Ulli Vilsmaier, Henrik von Wehrden, Paivi Abernethy, Christopher D. Ives, Nicolas W. Jager, and Daniel J. Lang. 2017. "Leverage Points for Sustainability Transformation." *Ambio* 46 (1): 30–39. https://doi.org/10.1007/s13280-016-0800-y.

Acuto, Michele, Susan Parnell, and Karen C. Seto. 2018. "Building a Global Urban Science." *Nature Sustainability* 1 (1): 2–4. https://doi.org/10.1038/s41893-017-0013-9.

Adamson, Joni, and Michael Davis, eds. 2016. *Humanities for the Environment: Integrating Knowledge, Forging New Constellations of Practice*. London: Routledge.

Boone, Christopher G. 2015. "On Hope and Agency in Sustainability: Lessons from Arizona State University." *Journal of Sustainability Education* 10 (November). http://www.susted.com/wordpress/content/on-hope-and-agency-in-sustainability-lessons-from-arizona-state-university_2015_11/.

Boone, Christopher G., Charles L. Redman, Hilda Blanco, Dagmar Haase, Jennifer Koch, Shuaib Lwasa, Harini Nagendra, Stephan Pauleit, Steward T.A. Pickett, Karen C. Seto, and Makoto Yokohari. 2014. "Reconceptualizing Land for Sustainable Urbanity." In *Rethinking Global Land Use in an Urban Era*, edited by Karen C. Seto and Anette Reenberg, 313–30. Cambridge, MA: The MIT Press.

Boone, Christopher G., Steward T.A. Pickett, Gabriele Bammer, Kamal Bawa, Jennifer A. Dunne, Iain J. Gordon, David Hart, Jessica Hellmann, Alison Miller, Mark New, Jean P. Ometto, Ken Taylor, Gabriele Wendorf, Arun Agrawal, Paul Bertsch, Colin Campbell, Paul Dodd, Anthony Janetos, and Hein Mallee. 2020. "Preparing Interdisciplinary Leadership for a Sustainable Future." *Sustainability Science* 15: 1723–33. https://doi.org/10.1007/s11625-020-00823-9.

Crow, Michael M., and William B. Dabars. 2020. *The Fifth Wave: The Evolution of American Higher Education*. Baltimore: Johns Hopkins University Press.

Davidow, Bill, and Michael S. Malone. 2015. "How 'Virtual Water' Can Help Ease California's Drought." *The Wall Street Journal* (Opinion), March 20, 2015. http://www.wsj.com/articles/bill-davidow-and-michael-malone-how-virtual-water-can-help-ease-californias-drought-1426891721.

Gordon, I.J., K. Bawa, G. Bammer, C. Boone, J. Dunne, D. Hart, J. Hellmann, A. Miller, M. New, J. Ometto, S. Pickett, G. Wendorf, A. Agrawal, P. Bertsch, C.D. Campbell, P. Dodd, A. Janetos, H. Mallee, and K. Taylor. 2019. "Forging Future Organizational Leaders for Sustainability Science." *Nature Sustainability* 2 (July): 647–49. https://doi.org/10.1038/s41893-019-0357-4.

Grenni, Sara, Katriina Soini, and Lummina Geertruida Horlings. 2020. "The Inner Dimension of Sustainability Transformation: How Sense of Place and Values Can Support Sustainable Place-Shaping." *Sustainability Science* 15 (2): 411–22. https://doi.org/10.1007/s11625-019-00743-3.

Horcea-Milcu, Andra-Ioana, David J. Abson, Cristina I. Apetrei, Ioana Alexandra Duse, Rebecca Freeth, Maraja Riechers, David P.M. Lam, Christian Dorninger, and Daniel J. Lang. 2019. "Values in Transformational Sustainability Science: Four Perspectives for Change." *Sustainability Science* 14 (5): 1425–37. https://doi.org/10.1007/s11625-019-00656-1.

Kapuscinski, Anne R., Arun Agrawal, Christopher Boone, Erin Bromaghim, Garrick E. Louis, and Dorceta E. Taylor. 2020. *Strengthening Sustainability Programs and Curricula in Higher Education*. Washington, DC: The National Academies Press.

Klinsky, Sonja, and Harald Winkler. 2014. "Equity, Sustainable Development and Climate Policy." *Climate Policy* 14 (1): 1–7. https://doi.org/10.1080/14693062.2014.859352.

Klinsky, Sonja, Timmons Roberts, Saleemul Huq, Chukwumerije Okereke, Peter Newell, Peter Dauvergne, Karen O'Brien, Heike Schroeder, Petra Tschakert, Jennifer Clapp, Margaret Keck, Frank Biermann, Diana Liverman, Joyeeta Gupta, Atiq Rahman, Dirk Messner, David Pellow, and Steffen Bauer. 2017. "Why Equity Is Fundamental in Climate Change Policy Research." *Global Environmental Change* 44 (May): 170–73. https://doi.org/http://dx.doi.org/10.1016/j.gloenvcha.2016.08.002.

Matson, Pamela, William C. Clark, and Krister Andersson. 2016. *Pursuing Sustainability: A Guide to the Science and Practice*. Princeton: Princeton University Press.

McPhearson, Timon, David M. Iwaniec, and Xuemei Bai. 2016. "Positive Visions for Guiding Urban Transformations toward Sustainable Futures." *Current Opinion in Environmental Sustainability* 22 (October): 33–40. https://doi.org/10.1016/j.cosust.2017.04.004.

Mi, Zhifu, Dabo Guan, Zhu Liu, Jingru Liu, Vincent Viguié, Neil Fromer, and Yutao Wang. 2019. "Cities: The Core of Climate Change Mitigation." *Journal of Cleaner Production* 207 (January): 582–89. https://doi.org/https://doi.org/10.1016/j.jclepro.2018.10.034.

Miller, Thaddeus R., Arnim Wiek, Daniel Sarewitz, John Robinson, Lennart Olsson, David Kriebel, and Derk Loorbach. 2014. "The Future of Sustainability Science: A Solutions-Oriented Research Agenda." *Sustainability Science* 9 (2): 239–46. https://doi.org/10.1007/s11625-013-0224-6.

Newman, Peter, and Isabella Jennings. 2008. *Cities as Sustainable Ecosystems: Principles and Practices*. Washington, DC: Island Press.

Noetel, Michael, Shantell Griffith, Oscar Delaney, Taren Sanders, Philip Parker, Borja del Pozo Cruz, and Chris Lonsdale. 2020. "Video Improves Learning in Higher Education: A Systematic Review." *PsyArXiv* (May). https://doi.org/10.31234/osf.io/kynez.

NPR. 2015. "Saudi Hay Farm in Arizona Tests State's Supply of Groundwater." November 2, 2015. http://www.npr.org/sections/thesalt/2015/11/02/453885642/saudi-hay-farm-in-arizona-tests-states-supply-of-groundwater.

Ostrom, Elinor, Marco A. Janssen, and John M. Anderies. 2007. "Going Beyond Panaceas." *Proceedings of the National Academy of Sciences* 104 (39): 15176–78. https://doi.org/10.1073/pnas.0701886104.

Ritchie, Hannah, Max Roser, and Esteban Ortiz-Ospina, and Jaiden Mispy. 2018. "Measuring Progress towards the Sustainable Development Goals." *SDG-Tracker.org*. https://sdg-tracker.org/.

Rogers, Zoey, and Elizabeth Bragg. 2012. "The Power of Connection: Sustainable Lifestyles and Sense of Place." *Ecopsychology* 4 (4): 307–18. https://doi.org/10.1089/eco.2012.0079.

Searchinger, Tim, and Ralph Heimlich. 2015. "Avoiding Bioenergy Competition for Food Crops and Land." Working Paper, Installment 9 of Creating a Sustainable Food Future. Washington, DC: World Resources Institute. https://www.wri.org/research/avoiding-bioenergy-competition-food-crops-and-land.

Seto, Karen C., Anette Reenberg, Christopher G. Boone, Michail Fragkias, Dagmar Hasse, Tobias Langanke, Peter Marcotullio, Darla K. Munroe, Branislav Olah, and David

Simon. 2012. "Urban Land Teleconnections and Sustainability." *Proceedings of the National Academy of Sciences* 109 (20): 7687–92. https://doi.org/10.1073/pnas.1117622109.

Solecki, William, Karen C. Seto, Deborah Balk, Anthony Bigio, Christopher G. Boone, Felix Creutzig, Michail Fragkias, Shuaib Lwasa, Peter Marcotullio, Patricia Romero-Lankao, and Timm Zwickel. 2015. "A Conceptual Framework for an Urban Areas Typology to Integrate Climate Change Mitigation and Adaptation." *Urban Climate* 14, Part 1 (December): 116–37. https://doi.org/10.1016/j.uclim.2015.07.001.

Tellman, Beth, Julia C. Bausch, Hallie Eakin, John M. Anderies, Marisa Mazari-Hiriart, David Manuel-Navarrete, and Charles L. Redman. 2018. "Adaptive Pathways and Coupled Infrastructure: Seven Centuries of Adaptation to Water Risk and the Production of Vulnerability in Mexico City." *Ecology and Society* 23 (1): 1. https://doi.org/10.5751/ES-09712-230101.

UN DESA (United Nations Department of Economic and Social Affairs). 2019. *World Urbanization Prospects: The 2018 Revision.* https://doi.org/10.18356/b9e995fe-en.

Weisman, Alan. 2007. *The World Without Us.* New York: Picador.

Wiek, Arnim, Lauren Withycombe, and Charles L. Redman. 2011. "Key Competencies in Sustainability: A Reference Framework for Academic Program Development." *Sustainability Science* 6 (2): 203–18. https://doi.org/10.1007/s11625-011-0132-6.

Xu, Zhenci, Sophia N. Chau, Xiuzhi Chen, Jian Zhang, Yingjie Li, Thomas Dietz, Jinyan Wang, Julie A. Winkler, Fan, Baorong Huang, Shuxin Li, Shaohua Wu, Anna Herzberger, Ying Tang, Dequ Hong, Yunkai Li, and Jianguo Liu. 2020. "Assessing Progress towards Sustainable Development over Space and Time." *Nature* 577 (7788): 74–78. https://doi.org/10.1038/s41586-019-1846-3.

4

How Trends in Public Higher Education Can Support Sustainability Education and Research

ROBERT H. JONES

HISTORICALLY, environmentally oriented teaching and research in higher education have been strongly cyclic, rising and falling in response to broad economic, social, and political forces. For example, since 1980, enrollments in natural resource fields at US universities have changed by as much as 50 percent within a decade (Sharik et al. 2015). The US Environmental Protection Agency's budget went up and down between 1970 and 2020, with a low of $3.0 million in 1981 and a high of $10.3 million in 2010 (EPA 2020). These fluctuations have hampered efforts to build the steady stream of trained researchers and professionals needed to meet the challenges of global environmental degradation. Current trends within public universities, however, provide hope for more stable and positive outcomes.

My voice in this chapter is influenced by forty-five years of higher education experience in natural resource programs, which started when I was a student and continued on in a postdoctoral position at the Savannah River Ecology Laboratory. As a faculty member in forestry or biology, I then honed my experience at Auburn University, Virginia Tech, West Virginia University, and Clemson University. My perspectives have been shaped by my research and teaching in both basic and applied ecology, directing twenty master's and PHD thesis projects, and my work as a higher education administrator for eighteen years, hiring and mentoring numerous ecology and environmental science faculty at three universities. I also practice what I believe in. On my own 72-hectare property in Virginia, I converted pastures dominated by non-native plants into native tall grass prairie, and I have placed the property into a conservation easement.

Public universities have become increasingly entrepreneurial, innovative, and masters of their own destiny. While universities are dependent on philanthropic donations, directed grants, industry funding, and community engagement, they are increasingly able to develop their own place-based strengths. Of the many changes universities are undergoing, four in particular provide avenues to build workforce capacity in the realm of sustainability: (1) changes in economic models to fund university operations; (2) shifts in general education pedagogy and learning approaches; (3) fundamental changes in curricula and philosophies associated with sustainability-related professional fields; and (4) changing demographics of college-eligible students. While these are not the only factors affecting sustainability programs, their impacts are already apparent and encouraging. From my perspective, sustainability resembles "flourishing" and emphasizes nurturing possibility for all human and non-human life on earth (Ehrenfeld and Hoffman 2013).

The Changing Funding Model

In post-World War II United States, the G.I. Bill and prevailing public opinion clearly signalled the importance of higher education as a path for a stronger economic future. Major public investments in higher education were made, primarily by state governments. Slowly at first, but then rapidly from the 1970s onward, state tax dollar support of public universities was eroded by rising health care costs and other competing interests. State government funding of higher education as a proportion of total appropriations declined between 1995 and 2019 (SHEF 2020, 11), while the student share of total university revenue grew, reaching 46 percent of total university revenues in 2019 (SHEF 2020, 10). This trend, coupled with rising costs to deliver high-quality teaching and research, has forced universities to not only raise tuition but also seek alternative avenues to grow revenues. Fortunately, new strategies to increase revenues have provided stability, and some of them—e.g., online learning, professional master's degrees offered on remote campuses, certificate programs, and other sorts of professional training—are not only major sources of revenues but also new spaces for innovation. Faculty in public universities have not simply followed this trend; they have been drivers of it. Their involvement has been incentivized through access to salary supplements and guarantees of intellectual property rights for certain kinds of discovery, enabled in particular by the US government's Act to Amend the Patent and Trademark Laws of 1980, also known as the Bayh-Dole Act, which provided universities and faculty with clearly defined ownership of intellectual property (IP) developed during research conducted at universities, as well as access to revenues that the faculty and university-owned IP may generate.

External markets—e.g., potential students—are strong for learning programs that feature novel thinking and desired credentials that improve job prospects. The interdisciplinary approaches needed to deal with the complexities of sustainability require entrepreneurial thinking and innovation, and both of these concepts are piquing potential student interest. Furthermore, many graduates from single discipline programs or with bachelor's degrees are now seeking additional credentials to advance within their sustainability-related career paths. By exploiting these two markets (i.e., students interested in interdisciplinary challenges and students interested in advanced credentials), universities are building a greater workforce capacity in sustainability.

Economic forces are also causing universities to pay closer attention to cost control through careful waste management, increased campus tree cover, commitments to carbon neutral footprints, use of fewer pesticides, and a host of additional sustainability-related practices. Combined with university efforts in sustainability research and education, this "green" approach has become strongly supported by university administrators, faculty, and students. This approach has also resulted in a proliferation of campus sustainability centres, each with education, research, and campus operations components. Many of these centres provide seed grants to stimulate innovation, engage students in unique learning opportunities, and make a real difference in cost control or energy savings. Ironically, the days in which administration and non-academic staff were perceived as barriers to environmentally sensitive business practices have been replaced by a new era in which the administration is a driver of innovation (Krizek et al. 2012). This trend reflects similar shifts in leadership and is playing out in all sectors of the US and world economies.

Changes in Pedagogy

Liberal education—an approach aimed at building critical thinking skills through exposure to different philosophies of learning and methods of problem solving—is a hallmark of North American higher education (AAC&U 2020). Since its widespread adoption in the second half of the twentieth century, liberal education has bifurcated into two major philosophies, one focused on the effectiveness of using simple course distribution models (i.e., choose at least one or two courses from group A, group B, etc.) and the other on interdisciplinary programs aimed at global-scale challenges, such as water quality and availability for all of humanity, global climate change, food security, and the future of alternative energy. The latter approach is being increasingly adopted, not only because it is perceived as a more powerful learning approach, but also because alumni and faculty wish to emphasize the uniqueness and strengths of their institution. This bodes well for sustainability topics because they are rich in the use of new technologies and in the emphasis on social and environmental sciences, which are highly interdisciplinary and well suited to meet liberal education learning outcomes.

Experiential learning—where tactile, visual, teamwork, and social interaction elements are core elements of the learning environment—has grown in importance in recent years, resulting in an explosive growth of study abroad programs, internships, service learning, undergraduate research, creative inquiry, and a host of other learning models. This trend also plays well to the strengths of sustainability education and research because hands-on field and laboratory experiences have long been core elements in sustainability disciplines (Domask 2007).

It is important to note that trends in liberal and experiential learning are directly affecting all disciplines in higher education, not just those related to sustainability. However, the trends may also have indirect impacts on the growth of a sustainability workforce by piquing student interest in sustainability majors, positively influencing public understanding of sustainability concepts, and influencing both political and economic decision making.

Changes in Professional Fields Related to Sustainability

In recent years, long-standing philosophical differences between different kinds of sustainability disciplines have been blurred, resulting in greater synergies between disciplines, attractive degree program alternatives that appeal to more students (Vincent et al. 2017), and better use of faculty resources. In the early to mid-twentieth century, sustainability-related disciplines were dominated by professional degree programs such as wildlife science, forestry, and civil engineering. These programs emphasized practical, business, and engineering elements of resource management, and not necessarily things such as biodiversity or social aspects. In more recent decades, new degree programs have emerged with more holistic and broader objectives and names that include words such as "sustainability," "conservation," or "environmental."

As the new programs developed, philosophies and learning objectives between them and the older, professional degree programs began to merge. This means that students can expect to learn core sustainability concepts in a broader selection of degrees while still being able to choose their specialization. Each of the two sets of programs brings strengths. The business and management tools highlighted in the professional degree programs are seen as important to the students in the more conservation-oriented programs. And some of the broader philosophical and social science elements of the conservation-oriented degrees are becoming critically important for professional degree holders, helping to prepare them for jobs in an environment where the public is demanding more agency in the use of natural resources. Many individual courses satisfy requirements for more than one of these degree programs. The net effects of these trends are the following: (1) a broader array of options for students, which attracts more students into sustainability fields and helps to dampen enrollment fluctuations; (2) more efficient use of faculty resources; and (3) more justification for hiring faculty with sustainability credentials.

Demographic Trends Among College-Eligible Students

Trends in the number of students that will be available to populate sustainability degree programs in the coming decade present both optimistic and cautionary notes. The number of US students enrolled in undergraduate degrees in all degree-granting post-secondary institutions is projected to grow by 2 percent between 2018 and 2029, from 16.6 to 17 million (Hussar et al. 2020, 127). The increases are primarily driven by minority and women participants. However, because minority participation in sustainability disciplines has been traditionally low, proactive efforts will be needed to capture this growth within the sustainability workforce. This is likely to happen because universities have strategic goals to create inclusive campuses with minority participation in all elements of the academy.

The recent surge in student interest in entrepreneurship and innovation—as discussed in Houshmand's chapter in this volume—offers an exciting opportunity to increase recruitment into sustainability disciplines and to innovate in and grow the sustainability economic space. Enrollments in classes focused on these topics and participation in student entrepreneurship clubs are on the rise. Men, women, and minority groups all show a strong interest.

The Interrelationships Among Trends

The four trends above are not isolated but are interconnected in important ways. For example, enrollments in sustainability majors drive the economic engine of the university, which in turn builds faculty capacity, creates attractive and effective sustainability campus centres, creates more experiential learning opportunities in sustainability, and justifies the use of sustainability as part of an overall liberal education strategy. Of all these factors, enrollment growth may be the most important driver, given the importance of tuition in the revenue stream of universities.

A Case Study

Clemson University provides a good example for the history of a changing emphasis in sustainability science and education, in particular, how stability has emerged as an outcome of the convergence of natural resource and sustainability professions and the creation of a single interdisciplinary academic unit. Like all land-grant institutions, Clemson was founded on principles of sustainability. A high point came in the 1930s, when visionary Clemson leaders purchased over 11,000 hectares of degraded land adjacent to the campus and conducted a massive restoration operation while growing programs in forestry and wildlife science. These foundations, combined with strong engineering and agricultural programs, set the stage for strength and impact in sustainability sciences. Yet the number of students majoring in natural resources has cycled up and down since the 1950s, with peak periods in 1963–64, 1976–79, 1999–2000, and 2014–19. High points have been more than two times the low points in the intervening valleys. The foundation for stability began in 1974, when the forestry degree was changed into a forest resource management degree. A new undergraduate degree in environmental and natural resources emerged in 2004 and has grown steadily in popularity.

Meanwhile, financial stresses to the institution in the mid-1990s and mid-2000s resulted in a major reorganization of colleges and departments. An additional college reorganization took place in 2015, not as a result of budget cuts, but rather to better align academic programming with future opportunities in research, learning, and outreach. From this crucible of frequent change, a new Department of Forestry and Environmental Conservation emerged that was able to leverage new university funding models, new emphases in active learning, changes in natural resource professions, and shifts in student demographics, including the steady growth of undergraduate women enrolled at Clemson, from 7,643 (46.1 percent of the student body) in 2012 to 10,313 (49.4 percent) in 2020. Greater stability has been achieved. For example, a slow collapse in forestry faculty and student enrollment was buffered by the emergence of the environmental and natural resources degree, and by the capacity of the faculty to spread teaching and research across a much more interdisciplinary group of scholars. Clemson renewed its emphasis on forestry faculty hiring, but with an eye toward achieving the buffering effect of having multiple flavours of degree programs with a sustainability core. Although reorganizations are costly and can have negative effects on faculty and staff productivity, funded research in sustainability has remained steady, holding its primacy at the university despite major increases in research funding for health, engineering, and other technical fields. Like many other public institutions, Clemson has added endowed faculty positions in sustainability, increased its emphasis on recycling and the reduction of energy use, and committed to reinventing general education requirements. These changes are converging to provide a bright future for sustainability science and education.

Conclusion

Sustainability education and science at public universities have been highly cyclic, resulting in fits and starts that disrupt the creation of a strong workforce and the stability and visibility of research programs. But signs of change at public universities—especially land-grant universities, where environmental problem solving comfortably fits the university's mission—are very encouraging. The higher education ecosystem now has components that favour stability, especially the rise of multiple degree options that mitigate risks of the ebb and flow of student enrollment and the size of the faculty. There are new elements in place to not only stabilize but also grow sustainability learning and discovery. These include economic incentives that reward individual faculty members, academic departments, and the whole institution when student enrollment increases. They also include teaching models that deliver interdisciplinary content and expand beyond the borders of the main campus. The strong commitments of universities to growing a more diverse faculty and student body will likely help address the long-standing challenge of low diversity in the sustainability-related professions and further strengthen the opportunity to grow. Finally, the growth of entrepreneurship and innovation as core academic subjects has great potential to open new spaces in sustainability science and education and ultimately help meet the challenge of global environmental degradation.

References

AAC&U (Association of American Colleges and Universities). 2020. *What Liberal Education Looks Like: What It Is, Who It's For, and Where It Happens*. Washington, DC: AAC&U. https://portal.criticalimpact.com/user/25043/image/whatlibedlookslike.pdf.

Domask, Joseph J. 2007. "Achieving Goals in Higher Education: An Experiential Approach to Sustainability Studies." *International Journal of Sustainability in Higher Education* 8 (1): 53–68. https://doi.org/10.1108/14676370710717599.

Ehrenfeld, John R., and Andrew J. Hoffman. 2013. *Flourishing: A Frank Conversation about Sustainability*. Stanford: Stanford University Press.

EPA (United States Environmental Protection Agency). 2020. "EPA's Budget and Spending." Last updated June 24, 2020. https://www.epa.gov/planandbudget/budget.

Hussar, Bill, Jijun Zhang, Sarah Hein, Ke Wang, Ashley Roberts, Jiashen Cui, Mary Smith, Farrah Bullock Mann, Amy Barmer, and Rita Dilig. 2020. *The Condition of Education 2020* (NCES 2020144). Department of Education. Washington, DC: National Center for Education Statistics. https://nces.ed.gov/pubs2020/2020144.pdf.

Krizek, Kevin J., Dave Newport, James White, Alan R. Townsend. 2012. "Higher Education's Sustainability Imperative: How to Practically Respond?" *International Journal of Sustainability in Higher Education* 13, no. 1 (January): 19–33. https://doi.org/10.1108/14676371211190281.

Sharik, Terry L., Robert J. Lilieholm, Wanda Lindquist, and William W. Richardson. 2015. "Undergraduate Enrollment in Natural Resource Programs in the United States: Trends, Drivers, and Implications for the Future of Natural Resource Professions." *Journal of Forestry* 113, no. 6 (November): 538–51. https://doi.org/10.5849/jof.14-146.

SHEF (State Higher Education Finance). 2020. *State Higher Education Finance (SHEF) Report. FY 2019*. https://shef.sheeo.org/wp-content/uploads/2020/04/SHEEO_SHEF_FY19_Report.pdf.

United States. 1980. H.R. 6933 – 96th Congress (1979–1980): An Act to Amend the Patent and Trademark Laws. Pub. L. No. 96-517. December 12, 1980. https://www.congress.gov/bill/96th-congress/house-bill/6933.

Vincent, Shirley, Sumedha Rao, Qiyuan Fu, Katt Gu, Xiao Huang, Kaitlyn Lindaman, Elishiva Mittleman, Kien Nguyen, Rachael Rosenstein, and Young-Jun Suh. 2017. *Scope of Interdisciplinary Environmental, Sustainability, and Energy Baccalaureate and Graduate Education in the United States*. Washington, DC: National Council for Science and the Environment.

II | Skill Sets or Research Capabilities Needed for Sustainability Education

5

Sustainability Education at US and Canadian Tribal Colleges
Its Goals and Implementations, and the Role of Mathematics

ROBERT E. MEGGINSON

MANY PEOPLE OF MY GENERATION will remember the "Crying Indian" commercial (Keep America Beautiful 1971),[1] in which an actor, dressed as a traditional American Indian,[2] navigates a canoe through trash floating in a polluted river and then lands on a refuse-littered shore, only to have a bag of trash tossed from a car break open at his feet.[3] The final image of the commercial, where the Indian has slowly turned to the camera with a tear running down his cheek, is one of the most powerful in the history of advertising, and it helped earn the commercial the number fifty spot on the Ad Age list of the top one hundred advertising campaigns of the twentieth century (Ad Age 1999).

In the areas in which it ran, this commercial and its final image helped cement in people's minds the notion that North American Indigenous Peoples are keepers of the Earth, who have learned how to live in sustainable harmony with the environment and particularly value the preservation of our planet's natural resources for the use and enjoyment of future generations. Of course, this is a stereotype when applied to individuals, and more than once I have had to say, "Hey, pick that up!" when one of my American Indian students in a summer program's outdoor project has littered something onto the ground.⁴ However, many North American Indigenous societies do place a strong emphasis on the sustainability of our environment and our traditional lifestyles and economies, and, where tribes are land based, such as on Canadian reserves and US reservations, recognize the importance of the sustainable development of tribal natural resources, particularly in service to a sustainable tribal economy.

My own view of sustainability has come in part from my observations of the value of the sustainable existence that my Lakota grandfather (born in 1882) and his wife lived, and that I experienced with him during long visits as a youth, as well as from my work with other tribes dating back more than thirty years. In particular, I spent much of many summers in the 1990s on the reservation of the Turtle Mountain Band of Chippewa Indians (Ojibwe, although I am using their own preferred name for their Nation) as an adjunct mathematics faculty member at their tribal college, which helped me understand the value of sustainability education to its students and how mathematics can support that. During my six years as associate dean for undergraduate and graduate education for the University of Michigan's College of Literature, Science, and the Arts, the University's Program in the Environment (PitE) reported to me, and I benefitted greatly from the wisdom of others about the value to everyone from all backgrounds of sustaining what the Earth and our ancestors have provided us.

This chapter focuses on the role of the Canadian and US tribal colleges in supporting the sustainability goals of the peoples they serve. For the purposes of this chapter, I adopt a common definition of sustainability from the Brundtland report: "Sustainable development is development that meets the needs of the present without compromising the ability of future generations to meet their own needs" (WCED 1987, 43). It must be emphasized that for North American Indigenous Peoples, our understanding of the meaning of "the needs of the present" includes at its very core the importance of preserving the tenets and practices of our cultures that have in many cases developed over millennia. Those have proved to be adaptable to meet new challenges, provided care is taken to listen to our knowledge keepers about wise, tested, and practical ways to adapt while preserving that cultural core. Elders will always be at the heart of that, although tribal colleges have also proved their value as knowledge keepers and adapters (while having themselves also stressed the value of the wisdom and guidance of Elders).

A table that summarizes the results of a survey of how sustainability education is implemented in tribal colleges, Table 5.1, can be found at the end of this chapter. As an example, the Menominee Indian Tribe of Wisconsin, an Algonquian woodland people, has strong cultural ties to its reservation's forests and sustainably maintains them for both cultural and economic reasons (Onesmus 2007). Because of this, it is not surprising that the College of Menominee Nation, a two- and four-year college established and maintained by the tribe, offers a two-year associate degree in natural resources emphasizing sustainable forestry, but also addressing the sustainability of fisheries, wildlife, and water and soil resources; at the time the information was gathered for this chapter (2016–17), it also offered an associate degree in the general area of sustainable development (as well as a "technical diploma," a one-year certificate program, in sustainable residential building systems). These efforts are supported by the College of Menominee Nation Sustainable Development Institute (College of Menominee Nation n.d.), which is devoted to research, education, and outreach.

The term "tribal college" is commonly used, as it will be here, for an institution of higher education whose student body consists, in fact and by intent, primarily of Indigenous people of Canada or the US. This designation has historically been applied to two-year colleges, four-year baccalaureate institutions, and universities (so, e.g., Haskell Indian Nations University is often called a tribal college, even by its own representatives). The terms "tribal college or university" and "TCU" are sometimes used for such an institution, but the use of the term "tribal university" without the word "college" in it is rare.[5] Most were founded, and are now operated, by a single tribe[6] or a few culturally related tribes, and most draw the bulk of their student bodies from those tribes. However, many also have Indigenous people from other tribes, as well as non-Indigenous ones, among their students. In particular, Haskell Indian Nations University and Southwestern Indian Polytechnic Institute, both operated by the US Department of the Interior's Bureau of Indian Education rather than by individual tribes, accept students from all federally recognized US tribes. The list of seventeen Canadian and thirty-seven US tribal colleges in Table 5.1 was drawn from multiple sources: the American Indian Higher Education Consortium (through Tribal College Journal of American Indian Higher Education n.d.); White House Initiative on American Indian and Alaska Native Education (n.d.); and (particularly for Canadian institutions) the Wikipedia article "List of Tribal Colleges and Universities," with some pruning to remove tribal colleges no longer in operation (e.g., D-Q University in California) or that are not actually colleges but instead are educational service organizations (e.g., the Iohahi:io Akwesasne Education & Training Institute in Ontario, which facilitates some post-secondary education through arrangements with off-site public colleges and universities).

The information used in this study and reported in Table 5.1 was collected from online college catalogs and supplemented by additional information from college websites and other internet sources.[7] Reasonable care was taken to find the courses and curricula at each institution with sufficient emphasis on sustainability to be classified primarily as sustainability education. However, the cultural value placed on sustainability is great enough in most of the peoples served by tribal colleges that sustainability is almost certainly interwoven into some courses and curricula in ways not evident from their catalog descriptions. This was certainly true at the tribal college at which I spent much time as a summer adjunct mathematics faculty member in the 1990s.

With no notable exceptions, the sustainability education offered at the fifty-four institutions under examination focuses on some type of sustainability that is a blend of four of the six definitions of that term given by Becky Brown and colleagues (Brown et al. 1987), namely, sustainable biological resource use, sustainable agriculture, sustainable society and economy, and sustainable development. In the Brown et al. paper, that last term is given a meaning somewhat different from the familiar one from the Brundtland report (WCED 1987, 43) that appeared the same year: "Sustainable development is development that meets the needs of the present without compromising the ability of future generations to meet their own needs." However, for the current discussion, the difference is not particularly important, and in fact the Brundtland report's definition is closer to what is intended here.

As already mentioned, in 2016–17, the College of Menominee Nation offered an associate degree in sustainable development that examined different dimensions of sustainability in more generality than the sustainable use of the Menominee forests, including sustainability's economic, technological, and behavioural aspects. With that exception, the sustainability education offered at the tribal colleges tends to focus fairly narrowly on sustainable use of the specific natural resources available to the tribes served by the colleges.[8]

Only five of the seventeen tribal colleges in Canada offer sustainability education (as such, but again, always keep in mind that sustainability may be interwoven into courses at tribal colleges in ways that are not externally obvious). One, First Nations University of Canada, has a Bachelor of Arts in Resource and Environmental Studies (requiring the first sixty hours to be completed at the Saskatchewan Polytechnic Woodland Campus), as well as a Bachelor of Science in Indigenous Environmental Science. The goals of those programs include education in sustainable management of Canadian resources, with some focus on Saskatchewan. One other institution, Yellowhead Tribal College, has a three-year advanced diploma in Indigenous Environmental Stewardship and Reclamation focusing on sustainable resource management, while three others have two-year diplomas in roughly equivalent areas. None of the other colleges have courses that are specifically described as having a sustainability focus.

Of the thirty-seven US tribal colleges in Table 5.1, only three have no course offerings specifically about sustainability. Interestingly, and perhaps significantly, those are the three tribal colleges in the state of Oklahoma, in which all tribes had their claims to tribally held reservation lands extinguished in a series of federal acts between 1887 and 1906.[9] Although sustainable economic development through tribal businesses and industry is still a major concern of the Oklahoma tribes, and one sees hints of this, for example, in some of the Pawnee Nation College courses in management, marketing, and business, sustainable agriculture and natural resource management is not such an important issue for tribes that have no land held in common to farm or natural resources to manage.

Of the other thirty-four colleges, six have no sustainability curricula leading to certificates or degrees, but they do have one or more courses with an explicit sustainability emphasis. An interesting example is the course in sustainable northern building construction taught at Iḷisaġvik College on Alaska's North Slope, possibly on a one-time basis. This is a premier example of a sustainability education offering designed with the specific needs of its audience in mind. This leaves twenty-eight US tribal colleges that have sustainability curricula leading to certificates or degrees, and all have at least associate degrees, with nineteen having that as the highest offering and eight of the others instead topping out with a bachelor's degree. The remaining college, Sitting Bull College, has a graduate degree emphasizing sustainability, its Master of Science in Environmental Science. The program includes a solid collection of mandatory and optional resource management courses addressing the specific resources that the Hunkpapa Lakota people, on whose reservation the campuses of the college are located, need to sustain.

One inescapable conclusion to be drawn from the descriptions of these tribal college programs is that sustainability education at these institutions almost always focuses sharply on the needs of the particular tribes served by the colleges, which is in line with, as stated by the American Indian Higher Education Consortium (n.d.), the founding of these institutions "for a specific purpose: to provide higher education opportunities to American Indians through programs that are locally and culturally based, holistic, and supportive."[10] This must be kept in mind when outside entities, such as majority-serving institutions of higher education, wish to engage with tribal colleges on issues involving sustainability education (or any other issues, for that matter). It should also be kept in mind that sustainability issues can be intertwined with conflicting economic ones for some tribes, such as the reliance of many Native Alaskans on fossil fuel revenues, and the coal mining operations on the Navajo reservation. This introduces complexities in conversations with some tribes about sustainability, which requires people from an external institution seeking a partnership to have knowledge of, experiences with, and full respect for those complexities, as well as the ability to frame and discuss them sensitively.[11]

Sustainability, as a strong interdisciplinary pursuit, borrows much from both the natural and social sciences, particularly as it is practiced by Indigenous Peoples interested in sustaining their natural resources. It is often said that mathematics is the language of science,[12] and the theory and practical applications of modern sustainability science make much use of that language.[13] Mathematics teaches students to understand scale, use conversion factors, read and create graphs, and make absolute and relative comparisons. Quantitative reasoning skills make it possible to understand ratios, scientific notation, and linear and exponential growth, and to speak with precision to the limitations and uncertainties of knowledge. All of these skills are critical in the sustainability sciences (Barwell 2018).

Consequently, students in a sustainability curriculum need to take enough mathematics to understand the underpinnings of their subject and be able to do the required applied mathematics. I will list here my own suggestions for basic mathematics requirements for degrees in sustainable management of natural resources, the most common type of sustainability degree program found at tribal colleges, and follow that with a quick look at how closely tribal college sustainability programs match up with these recommendations.

- *One-year programs, including certificates:* College algebra. Students with any certification of expertise in sustainability science should be able to interpret information given to them graphically, numerically (particularly in tables), and algebraically (particularly as formulas), and then be able to convert such information between those forms as needed. Students should also be able to reinterpret for others, particularly for a general audience, information given to them in any of these forms. Linear equations should be studied in enough depth to be able to create simple linear models, e.g., a formula predicting the height of a tree in future years, based on its known height in past years. Students should also be able to recognize from data when a linearity assumption does not appear to be valid. All of that, as well as much other useful mathematics not covered in this paragraph, is the stuff of college algebra.
- *Two-year programs, including associate degrees:* Precalculus at a minimum, and, ideally, basic applied differential and integral calculus. The precalculus course should include a presentation of trigonometric functions as tools for modelling periodic behaviour, not just as ways to relate parts of triangles to each other. Modern precalculus courses often include an introduction to fitting curves to data using technology, which can prove useful to resource managers. Calculus should at least be

available to students who would be up to a challenge, in particular due to the knowledge gained about rates of change and their applications.[14]

Tribal colleges that have evolved into baccalaureate and graduate institutions should also aim to have the following mathematics requirements for more advanced degrees.

- *Bachelor's degree programs:* Differential and integral calculus through at least a basic introduction to systems of ordinary differential equations (ODEs). As one example of the usefulness of this in sustainable management of wildlife, the standard predator–prey model is given by a pair of first-order nonlinear ODEs.
- *Graduate programs:* A standard course in partial differential equations (PDEs), but including an elementary introduction to dynamical systems and critical transitions at least extensive enough to understand hysteresis. The behaviour of natural systems in which many factors interact can often be described by systems of PDEs. Also, an understanding of critical transitions can be of value to a reservation water resource specialist trying to figure out why a slight increase in phosphate runoff into a fishing lake can suddenly lead to dramatic eutrophication, which does not go away when the phosphate level in the lake is reduced to its pre-eutrophication level.[15]

A few disclaimers and confessions are necessary before looking at how closely the tribal colleges are approaching these suggested requirements. First of all, while the counts to follow are based in good part on the data in the accompanying table at the end of this chapter, in a few cases I used some judgment in deciding that the names of the mathematics courses required for a particular program did not tell the full story about the strength of the requirement.[16] Also, the creative naming of some mathematics courses, along with the lack of syllabi or other documentation from which actual course content could be divined, means that it is at least technically possible that some courses that do satisfy my suggested requirements were not given the benefit of the doubt. Furthermore, that creative naming phenomenon is particularly prevalent for the mathematics courses required for one-year certificate programs, so I admit to giving up on my effort to determine whether courses for those programs effectively satisfy the college algebra requirement. Finally, as with all colleges, the catalogs and other educational documentation of the tribal colleges are always works in progress, and Table 5.1 is only a snapshot at one point in time, in most cases the 2016–17 academic year. With all of that in mind, the numbers for two-year and more advanced degree programs for the colleges can be found in that table.

Of the twenty-seven two-year degree programs for which I was able to determine whether they satisfy the suggested precalculus requirement, eight do, and two of the eight exceed that minimum by mandating a calculus course. Another twelve require college algebra instead, which puts a student who wishes to transfer to a four-year institution (or continue at the same institution in a bachelor's degree program) within one course of starting the calculus sequence, not an insurmountable problem for such students. Moreover, all but one of those twelve programs are at institutions offering precalculus, so, with that single exception, the course is available for students who need it to round out a solid associate degree or to prepare them better for a bachelor's program. Only seven of the twenty-seven have basic mathematics requirements below college algebra. While those seven should consider stiffening their requirements, the mathematics that students take for a two-year sustainability degree at most tribal colleges is no more than one notch below precalculus.

In contrast, of the ten bachelor's degrees offered at the tribal colleges that have a sustainability focus, only two meet the suggested mathematics requirement of differential and integral calculus through basic ODE systems, and only one additional degree requires any calculus at all. Moreover, only one more requires precalculus, although another may effectively have that requirement through a combination of courses. The other five degree programs require only college algebra.

The single master's program with a sustainability emphasis mentioned in the table, at Sitting Bull College, requires only college algebra, and that may be sufficient, for the courses in that interesting degree program are exceptionally carefully structured to target the particular ecological and environmental areas of most relevance to the Hunkpapa Lakota people. However, for future expansion of the program, and for planning for graduate programs at other tribal colleges, more could be required, particularly because Sitting Bull College does offer mathematics at much higher levels.

Some of these tallies may seem to cast an unfavourable light on the overall efforts of tribal colleges to provide appropriate mathematics to students in their sustainability curricula. I would urge the reader not to be too quick to place the blame on the colleges. To address this matter, I would first plead that the two nations and their provinces, territories, and states in which these institutions operate help the colleges obtain the resources to make curricular upgrading possible. From my own experience as an adjunct tribal college faculty member, I can testify that these colleges do a remarkable job of taking students as they come to them, sometimes with mathematics preparations several grade levels below what would be desirable for their ultimate educational goals, and of bringing them farther along than the students themselves might have thought possible. However, the resulting developmental math teaching loads at these colleges often leave little time and other educational resources to devote to advanced mathematics courses that serve only a handful of students, and the long remedial paths required to get more students ready to take those advanced courses can be intimidating for both student and teacher. This is a complicated problem that has its roots partly in under-resourced K–12 programs.

Conclusion

In closing, I would like to revisit the claim that introduced the preceding discussion of mathematics, that "sustainability, as a strong interdisciplinary pursuit, borrows much from both the natural and social sciences," because it is incomplete. As both a field of knowledge and a cultural value within Indigenous communities, sustainability also has a strongly philosophical side to it. In particular, sustainability is an important aspect of the Seventh Generation Principle, found by essentially that name in many North American Indigenous cultures and implicitly in other Indigenous societies worldwide: every decision we make, every action we take, should be done in full consideration of its impact on our descendants seven generations into the future.[17] (Although "seven generations into the future" is interpreted literally by some North American Indigenous cultures, for this discussion it may be thought of as a metaphoric reference to all future generations.)

It is theoretically possible to teach sustainability as a collection of facts and techniques taken from the natural sciences and related fields, supported by the social science needed to convince human beings and governmental units of its value for people already living. Sustainability education at the tribal colleges generally goes beyond that by implicitly, and often explicitly, incorporating the philosophy of the Seventh Generation Principle to provide an ethical basis for arguing the importance for generations yet unborn of our practicing sustainability. This aspect of tribal college sustainability education could usefully inform the teaching of sustainability in higher education more generally, although I would recommend being explicit about the ethical principle involved when teaching students from cultures that do not stress it. To this end, it can be helpful to find an eloquent statement supporting the underlying principle of concern about impacts of our current practices on future generations, which can come from anyone from any culture, provided the instructor finds it personally compelling. It can then be distributed to students for a classroom discussion. I will end with such a statement that plays that role for me, by Chief Oren Lyons, Faithkeeper of the Turtle Clan of the Seneca Nation of the Haudenosaunee (Iroquois) Confederacy and member of the Onondaga Council of Chiefs of the Haudenosaunee. This is taken from an essay on the Haudenosaunee perspective on sustainability that appeared in 1980. Importantly, Chief Lyons' words are addressed not just to American Indians, but to all people of our mother, the Earth.

> We are looking ahead, as is one of the first mandates given to us as chiefs, to make sure and to make every decision that we make relative to the welfare and well-being of the seventh generation to come, and that is the basis by which we make decisions in council. We consider: will this be to the benefit of the seventh generation? That is a guideline. We have watched various forms of governments, we have watched internationally the development of industry. We have watched within our own nations and territories the exploitation of not only the people but the resources without regard to the seventh generation to come. We

are facing together, you and I, your people and my people, your children and my children, we are facing together a very bleak future. There seems to be at this point very little consideration, minimum consideration, for what is to occur, the exploitation of wealth, blood, and the guts of our mother, the earth. Without the earth, without your mother, you could not be sitting here; without the sun, you would not be here...

Today belongs to us, tomorrow we'll give it to the children, but today is ours. You have the mandate, you have the responsibility. Take care of your people—not yourselves, your people. (Lyons 1980, 173–74)

TABLE 5.1: Tribal College and University Sustainability Education in Canada and the US as of Fall 2016

All data was gathered from course catalogs available online as of Fall 2016. As mentioned in the chapter, the cultural value placed on sustainability is great enough among most of the peoples served by tribal colleges that sustainability is almost certainly interwoven into some courses and curricula in ways not evident from their catalog descriptions. This list documents only the presence of sustainability curricula and courses explicitly described as such, or with descriptions that make apparent a sustainability emphasis. A number appearing in the "sustainability courses offered" column rather than just "yes" or "no" is a count of sustainability courses at an institution with no sustainability curricula but with a small but positive number of sustainability course offerings. It must be emphasized that this is a snapshot of sustainability at tribal colleges at a single point in time. Due to the volatility of course offerings at almost any institution of higher education, current information for any particular college in this list should be sought from that institution through an up-to-date catalog or direct contact.

Tribal college/ university, location of main campus	Sustainability courses offered	Sustainability degree programs	Focus	Mathematics required	Most advanced mathematics course at institution
Canada					
Alberta					
Blue Quills First Nations College, St. Paul	No				Basic statistics
Maskwacis Cultural College, Maskwacis	No				Business mathematics
Old Sun Community College, Siksika	No				Precalculus
Red Crow Community College,[18] Cardston	See note	See note	See note	See note	See note
Yellowhead Tribal College, Edmonton	Yes	1 year cert; 2, 3 year diplomas	Indigenous environmental stewardship and reclamation	Precalculus	Calculus
British Columbia					
Native Education College, Vancouver	No				College algebra
Nicola Valley Institute of Technology, Merritt	Yes	1 year cert, 2 year diploma	Environmental resource sustainability	Precalculus	Precalculus
Nunavut					
Nunavut Arctic College, Iqaluit	Yes	1 year cert, 2 year diploma	Environmental resource sustainability	Mathematical Foundations	Applied and business mathematics courses
Ontario					
Anishinabek Educational Institute, North Bay	Yes	2 year diploma	Forestry and renewable energy	Applied Mathematics in Natural Resource Sciences	Applied and business mathematics courses
First Nations Technical Institute, Tyendinaga	No				

Tribal college/ university, location of main campus	Sustainability courses offered	Sustainability degree programs	Focus	Mathematics required	Most advanced mathematics course at institution
Ontario (cont.)					
Kenjgewin Teg Educational Institute, M'Chigeeng	No				
Negahneewin College, Thunder Bay	No				
Oshki-Pimache-O-Win Education and Training Institute, Thunder Bay	No				
Seven Generations Education Institute,[19] Fort Frances	No				
Shingwauk Kinoomaage Gamig, Garden River	No				
Six Nations Polytechnic, Ohsweken	No				
Saskatchewan					
First Nations University of Canada,[20] Regina	Yes	BS	Environmental resource sustainability	Calculus I	Graduate level courses, through University of Regina, with which the program collaborates.
United States					
Alaska					
Iḷisaġvik College,[21] Barrow	1		Sustainable northern shelter construction		

90 *Sustainability Education at US and Canadian Tribal Colleges*

Tribal college/ university, location of main campus	Sustainability courses offered	Sustainability degree programs	Focus	Mathematics required	Most advanced mathematics course at institution
Arizona					
Diné College,[22] Tsaile	Yes	AS	Environmental science, preparatory for transfer in resource management, wildlife management, and other sustainability degree programs	Precalculus	Ordinary differential equations, linear algebra
Tohono O'odham Community College, Sells	Yes	AS	Implicit component of Tohono O'odham Agriculture and Natural Resources concentration	Calculus I	At least Calculus III
Kansas					
Haskell Indian Nations University, Lawrence	Yes	BS	Environmental science concentrations include sustainability as a core concept	College algebra	Linear algebra, differential equations
Michigan					
Bay Mills Community College, Brimley	1		Natural resource management, emphasizing forests		Analytic trigonometry
Keweenaw Bay Ojibwa Community College, Baraga	Yes		Some environmental science courses have a partial emphasis on sustainable management of tribal forest and water resources	College algebra	

Tribal college/ university, location of main campus	Sustainability courses offered	Sustainability degree programs	Focus	Mathematics required	Most advanced mathematics course at institution
Michigan (cont.)					
Saginaw Chippewa Tribal College, Mount Pleasant	2		Sustainability issues are addressed in a basic introductory environmental science course, plus another Native American environmental issues course		
Minnesota					
Fond du Lac Tribal and Community College, Cloquet	Yes	AS	Environmental science; strong emphasis on sustainability	College algebra	Calculus II
Leech Lake Tribal College, Cass Lake	Yes	AS	Earth system science and forest ecology concentrations with some emphasis on sustainability	College algebra	Calculus II
Red Lake Nation College, Red Lake	Yes	AS	Environmental science; some sustainability content	Concepts in mathematics	College algebra
White Earth Tribal and Community College,[23] Mahnomen	Yes	AA	Environmental science; some sustainability content		College algebra
Montana					
Aaniiih Nakoda College, Harlem	Yes	AS	Environmental science	College algebra	Calculus
Blackfeet Community College, Browning	Yes	AS	With careful course selection, an AS in General Studies—Math & Science can get at least a minor focus on sustainability	Probability & linear math	Calculus II

Tribal college/ university, location of main campus	Sustainability courses offered	Sustainability degree programs	Focus	Mathematics required	Most advanced mathematics course at institution
Montana (cont.)					
Chief Dull Knife College, Lame Deer	4		Sustainable forest and soil management, social issues, issues with mining		Calculus II
Fort Peck Community College, Poplar	Yes	AS	Environmental science	College algebra	Differential equations
Little Big Horn College, Crow Agency	Yes	AS	Two concentrations, Natural Resrcs/ Env Science and Tribal Natural Resrcs/Env Science	Precalculus or introductory statistics	Calculus II
Salish Kootenai College, Pablo	Yes	AS, BS	Multiple degrees emphasizing sustainability, e.g., in wildlife and fisheries, forestry, hydrology	Calculus II for BS	Calculus III, Mathematical Modeling
Stone Child College, Box Elder	Yes	AA	Water quality	College algebra	Calculus II
Nebraska					
Little Priest Tribal College, Winnebago	Yes	AS	Indigenous science, with sustainability emphasis	College algebra	College algebra
Nebraska Indian Community College, Macy	Yes	AS	General Science/ Natural Resources emphasis has a strong focus on sustainable management of forests, range, soil and water, and wildlife	Calculus I	Calculus I

Tribal college/ university, location of main campus	Sustainability courses offered	Sustainability degree programs	Focus	Mathematics required	Most advanced mathematics course at institution
New Mexico					
Institute of American Indian Arts, Santa Fe	1		Sustainability, Innovation & Entrepreneurship, focuses on sustainable business practices for competitive advantage		
Navajo Technical University,[24] Crownpoint	Yes	1 year cert, AAS, BS	Environmental science and natural resources	For BS: Precalculus	Differential equations, linear algebra, numerical analysis
Southwestern Indian Polytechnic Institute,[25] Albuquerque	Yes	AAS	Environmental science, natural resources management, some emphasis on soils	College algebra, plus trigonometry or statistics	Calculus II
North Dakota					
Cankdeska Cikana Community College, Fort Totten	Yes	AS	Environmental science, natural resources management	Intermediate algebra for environmental science, college algebra for natural resources management	Differential equations
Nueta Hidatsa Sahnish College, New Town	Yes	AS, AAS, BS	Environmental science	AS, BS: College algebra; AAS: any mathematics course	Differential equations
Sitting Bull College, Fort Yates	Yes	AS, BS, MS	Environmental science	AS: Intermediate algebra; BS, MS: College algebra	Differential equations

Tribal college/ university, location of main campus	Sustainability courses offered	Sustainability degree programs	Focus	Mathematics required	Most advanced mathematics course at institution
North Dakota (cont.)					
Turtle Mountain Community College, Belcourt	Yes	AS	Pre-environmental science; not as structured as most, but a sustainability emphasis can be designed into a program of study	College algebra	Differential equations
United Tribes Technical College, Bismarck	Yes	AAS, BS	Environmental science; not as focused on sustainability as most ES programs, but it is present in the core	AAS: College algebra; BS: Calculus I	Calculus III
Oklahoma					
College of the Muscogee Nation, Okmulgee	No				
Comanche Nation College,[26] Lawton	No				
Pawnee Nation College, Pawnee	No				
South Dakota					
Oglala Lakota College, Kyle	Yes	BS	Natural science, conservation biology track; some emphasis in earth science track	Trigonometry	Calculus III, linear algebra
Sinte Gleska University, Mission	Yes	AS, BS	Environmental science	AS: College algebra; BS: College algebra, statistics	Partial differential equations
Sisseton Wahpeton College, Sisseton	Yes	AA	Sustainable environmental studies	Intermediate algebra	Trigonometry

Tribal college/ university, location of main campus	Sustainability courses offered	Sustainability degree programs	Focus	Mathematics required	Most advanced mathematics course at institution
Washington					
Northwest Indian College,[27] Bellingham	Yes	AS, BS	Native environmental science	College algebra	Calculus III
Wisconsin					
College of Menominee Nation, Keshena	Yes	AAS	Sustainable development	College algebra foundations	Differential equations with linear algebra
Lac Courte Oreilles Ojibwa Community College, Hayward	Yes	AAS	Several natural resource management curricula	Intermediate algebra	Calculus I with analytic geometry

Notes

1. The commercial was a joint effort of the non-profit organizations Keep America Beautiful and the Advertising Council and was created by the advertising agency Marsteller Inc. (Advertising Council n.d.). Unfortunately, the video quality of existing renditions of that particular version of the famous commercial, including the one cited in this chapter, is poor. However, in 2014, a higher quality ad—a somewhat different version, in a different setting—created for cultural and educational purposes, was discovered and posted on YouTube through a joint research project of the Advertising Council and Chad Weidner of Utrecht University (Keep America Beautiful n.d.).

2. Since Keep America Beautiful and the Advertising Council are US-based organizations, the term I am using in the specific context of this commercial for a US person of Indigenous North American stock, whose ancestors arrived here during the last glaciation, is the one generally favoured by US Indigenous people of my generation when we are not using our specific tribal identifications. The term "Native American" is considered by some persons to be a synonym, and perhaps as preferable and more politically correct, but it should be recognized that some American Indians, such as the late, prominent American Indian activist Russell Means, detest it. Residents of Canada of the same ancestry, who are covered by Canada's *Indian Act* of 1876 are generally called First Nations (and often consider the term "Indian" offensive, particularly when used by non-First Nations people). The Inuit of Canada, another collection of Indigenous North American peoples, whose ancestors arrived later than the First Nations, are genetically distinct from the latter. Together with the mixed-blood Métis, the First Nations and Inuit Peoples are commonly referred to in Canada as Aboriginal. As with anyone on this planet, the best rule for what to call people is to find out what term they prefer, then use it.

3. The actor, Iron Eyes Cody (birth name Espera Oscar de Corti), long thought to be an American Indian himself, was actually the son of Sicilian immigrants, although that was not known outside his small hometown of Gueydan, Louisiana at the time the commercial was made. He had been portraying American Indians in films since the 1930s and had become famous by 1971. A reporter from *The New Orleans Times-Picayune* who investigated Cody's ancestry in 1996 discovered his Sicilian roots, although Cody continued to maintain he was American Indian until his death in 1999 (Crockett 2014).
4. Although that is rarely a soda bottle in my home state of Michigan, where they can be turned back in for a dime. I am a big fan of bottle and can deposit laws.
5. One US institution on the list to be examined, Fond du Lac Tribal and Community College, has a student body that, as of 2013, was about 64 percent white and, altogether, about 75 percent non-Indian (ProPublica n.d.). It is included in the list of tribal colleges in Table 5.1 because of its unique history and founding in collaboration with the Fond du Lac Band of Lake Superior Chippewa (Fond du Lac Tribal and Community College n.d.).
6. For lack of a better term, in this context "tribe" will sometimes be understood to mean a group of culturally related Aboriginal or American Indian people, including such groups as the Inuit, not necessarily bound together as a single political entity. This is arguably an abuse of language, but for the purposes here, no confusion should result.
7. One Canadian and one US institution originally in the table were ultimately removed from it and not included in this study because the security software on the computers used for site access characterized the college websites as high risk and issued strong warnings against visiting them.
8. This is closely related to what Charles Kidd (1992) has called the resource/environment and ecodevelopment roots of sustainability.
9. The classic, well-researched, and blood-boiling book on how this was done is Angie Debo's (1973) *And Still the Waters Run*, which originally appeared in 1940 and is still in print. This book has been credited by both Indian and non-Indian historians as causing a fundamental change in how American Indian history is written.
10. See Boyer (2015) for an in-depth look at the founding purposes and challenges of the tribal colleges.
11. For a discussion of complexities concerning climate change in interactions with Native Alaskans, see McNeeley and Huntington (2007), particularly 143–47.
12. This observation is not modern, but dates back at least to Galileo in *The Assayer* (Galilei 1623), a pioneering work on the importance of the scientific method. The following is the relevant passage from that essay, for which I must take responsibility for the accuracy of the translation from Italian: "Philosophy is written in this great book (the universe) which is constantly open before our eyes, but it cannot be understood unless we have first learned to understand the language and symbols in which it is written. That language is mathematics." The reference to philosophy is to natural philosophy, a concept from which our modern notion of the natural sciences is considered to have evolved.

13. More generally, modern mathematics is playing an ever-increasing role in both the natural and social sciences, and many curricula at many colleges could stand to pay more attention to that fact (Megginson 2007).
14. Knowing something about the interplay between rates of change and values of functions can help when someone is faced with an argument such as that once made by a US senator, in which he claimed that, because there is evidence that the temperature in the Arctic in 1935 was rising faster relative to its rise in other parts of the world than it was in 2003, therefore it must have been warmer in the Arctic in 1935 than it was in 2003 (Inhofe 2003, 182).
15. See Scheffer (2009) for an interesting introduction to these phenomena, accessible to someone who has taken the standard calculus sequence through differential equations.
16. That judgment did not always result in a lowered estimation. In one case, no precalculus course was required for a two-year program, but college algebra and the accompanying required courses brought the mathematics up to the precalculus level.
17. This particular statement of the Seventh Generation Principle is my own wording, as I have come to understand it from the wisdom of Elders and observations of its implementation, but a quick search with an internet browser on the various forms of the principle should provide convincing evidence that this captures the concept. The following passage from the Gayanashagowa, or Great Law of Peace, of the Haudenosaunee (Iroquois) Six Nations is generally accepted as its first formalization, stated here as it appears in the 1916 Parker version of the Gayanashagowa: "Look and listen for the welfare of the whole people and have always in view not only the present but also the coming generations, even those whose faces are yet beneath the surface of the ground—the unborn of the future Nation" (Parker 1916). The "Parker version" of the Gayanashagowa was based in part on the first English translation of the Gayanashagowa, by Seth Newhouse, Mohawk, in about 1880, and also on a document compiled by some of the Chiefs of the Six Nations Council in 1900. The time of origin of the Gayanashagowa is disputed and is variously given as 1390 or about 1450 to 1500, or possibly as early as about 1100. It was first recorded using wampum symbols.
18. Partners with University of Lethbridge (one hour away) and University of Calgary on its programs.
19. Education and training institute; does have articulation agreement for BA programs with Lakehead University.
20. FNU also hosts a transfer program in this area leading to a BA.
21. This was a one-time offering, but sustainable architecture may be an institutional interest.
22. Much emphasis on sustainability in food production, traditional food production, and development, with a strong emphasis on the Navajo Nation.
23. A bit thin on sustainability, but some is there.
24. The institution's mission statement emphasizes "environmental preservation and sustainable economic development"; see Navajo Technical University, "Mission, Vision, and Philosophy," accessed November 26, 2016, http://www.navajotech.edu/about/mission.

25. SIPI's strategic goals include "to better support the sustainability of tribes' fundamental needs"; see SIPI's "Mission and Vision," accessed March 22, 2022, https://www.sipi.edu/apps/pages/mission/vision.
26. Sustainability is one topic among many in a social problems course.
27. Sustainability less evident than in many other environmental science courses, particularly in AS program. It is likely embedded in seminars and internship in BS program.

References

Ad Age. 1999. "Ad Age Advertising Century: Top 100 Campaigns." March 29, 1999. http://adage.com/article/special-report-the-advertising-century/ad-age-advertising-century-top-100-advertising-campaigns/140150/.

Ad Council. n.d. "Pollution: Keep America Beautiful." The Classics. https://www.adcouncil.org/our-story/our-history/the-classics.

American Indian Higher Education Consortium. n.d. "About AIHEC." Accessed November 25, 2016. http://www.aihec.org/who-we-are/index.htm.

Barwell, Richard. 2018. "Some Thoughts on a Mathematics Education for Environmental Sustainability." In *The Philosophy of Mathematics Education Today*, edited by Paul Ernest, 145–60. ICME-13 Monographs. Cham, Switzerland: Springer. https://doi.org/10.1007/978-3-319-77760-3_9.

Boyer, Paul. 2015. *Capturing Education: Envisioning and Building the First Tribal Colleges*. Reprint, Pablo, MT: Salish Kootenai College Press.

Brown, Becky J., Mark E. Hanson, Diana M. Liverman, and Robert W. Meredith, Jr. 1987. "Global Sustainability: Toward Definition." *Environmental Management* 11, no. 6 (November): 713–19. https://doi.org/10.1007/BF01867238.

College of Menominee Nation. n.d. "College of Menominee Nation Sustainable Development Institute." Accessed November 25, 2016. http://sustainabledevelopmentinstitute.org.

Crockett, Zachary. 2014. "The True Story of 'The Crying Indian.'" *Priceonomics*. September 9, 2014. https://priceonomics.com/the-true-story-of-the-crying-indian/.

Debo, Angie. 1973. *And Still the Waters Run: The Betrayal of the Five Civilized Tribes*. Princeton: Princeton University Press.

Fond du Lac Tribal and Community College. n.d. "History." Accessed November 25, 2016. https://fdltcc.edu/admissions/about-us/history/.

Galilei, Galileo. 1623. *Il Saggiatore [The Assayer]*. Rome: Giacomo Mascardi.

Inhofe, James M. 2003. "The Science of Climate Change: Senate Floor Statement." In *The Global Warming Reader: A Century of Writing about Climate Change*, edited by Bill McKibben, 165–92. New York: Penguin Books.

Keep America Beautiful. 1971. "The Crying Indian—Full Commercial." YouTube video, 1:00. Posted April 30, 2007. https://www.youtube.com/watch?v=j7OHG7tHrNM.

Keep America Beautiful. n.d. "Crying Indian on Horseback—Iron Eyes Cody." YouTube video, 1:00. Posted March 5, 2014. https://www.youtube.com/watch?v=8_QGBWaD-A4.

Kidd, Charles V. 1992. "The Evolution of Sustainability." *Journal of Agricultural and Environmental Ethics* 5, no. 1 (March): 1–26. https://doi.org/10.1007/BF01965413.

Lyons, Oren. 1980. "An Iroquois Perspective." In *American Indian Environments: Ecological Issues in Native American History*, edited by Christopher Vecsey and Robert W. Venables, 171–74. Syracuse: Syracuse University Press.

McNeeley, Shannon, and Orville Huntington. 2007. "Postcards from the (Not So) Frozen North: Talking about Climate Change in Alaska." In *Creating a Climate for Change: Communicating Climate Change and Facilitating Social Change*, edited by Susanne C. Moser and Lisa Dilling, 139–52. Cambridge: Cambridge University Press.

Megginson, Robert. 2007. "The Mathematicization of Science." Lead-off editorial to 20th anniversary issue. *Winds of Change* 22, no. 1 (Winter).

Onesmus, Doris Karambu. 2007. "Sustainable Management of Forest by Menominee Tribe from Past to Present." *Native American Forestry*. University of Wisconsin-Stevens Point. Accessed November 25, 2016. http://www.uwsp.edu/forestry/StuJournals/Documents/NA/donesmus.pdf.

Parker, Arthur C. 1916. *The Constitution of the Five Nations*. Published as New York State Museum Bulletin, no. 184. Albany: The University of the State of New York.

ProPublica. n.d. "Fond du Lac Tribal and Community College." Accessed November 25, 2016. https://projects.propublica.org/colleges/schools/fond-du-lac-tribal-and-community-college.

Scheffer, Marten. 2009. *Critical Transitions in Nature and Society*. Princeton: Princeton University Press.

Tribal College Journal of American Indian Higher Education. n.d. "Tribal Colleges and Universities." *Tribal College Journal of American Indian Higher Education*. Accessed November 25, 2016. http://www.tribalcollegejournal.org/map-of-tribal-colleges.

White House Initiative on American Indian and Alaska Native Education. n.d. "Tribal Colleges and Universities." Accessed November 25, 2016. http://sites.ed.gov/whiaiane/tribes-tcus/tribal-colleges-and-universities.

Wikipedia. n.d. "List of Tribal Colleges and Universities." Accessed November 25, 2016. https://en.wikipedia.org/wiki/List_of_tribal_colleges_and_universities.

WCED (World Commission on Environment and Development). 1987. *Report of the World Commission on Environment and Development: Our Common Future*. Oxford: Oxford University Press.

6
Innovation
Connecting Markets and Money

VICKY J. SHARPE

I AM DRIVEN by a deep appreciation for the natural world. For a long time, I have sought approaches to protecting the environment in ways that make practical business sense and that provide economic returns to a broad swath of society—or what many recognize as sustainable development. Fundamentally, the Brundtland report definition works for me, i.e., "Sustainable development is development that meets the needs of the present without compromising the ability of future generations to meet their own needs" (WCED 1987, 43). However, interpretation and aspiration levels around the concept of sustainability vary widely across economies and geographies. Status quo thinking is not an option, with the risks of inaction becoming increasingly apparent. As the dominant species, we must look to how we govern ourselves, minimize destruction, and maximize benefits while ensuring social equity.

As a problem solver with a multidisciplinary educational background rooted in the biological sciences, I am driven to seek practical solutions. My exposure to the renowned Maurice Strong while I was a senior executive in the international subsidiary of a large energy company opened my eyes to the breadth of global opportunities to protect the environment, improve social equity, and strengthen economic performance, both at the project site and for the company. Since then, my exposure to the world of finance and investment, both public and private, through my board and advisory roles, has honed my views. I am now a crusader for clean capitalism.

My views on sustainability have been moulded further by my time as the founding president and CEO of Sustainable Development Technology Canada (SDTC), a federally funded foundation whose lofty mission is to act as the primary catalyst in building a sustainable development technology infrastructure in Canada. As a policy delivery agent, SDTC was created in 2001 as part of Canada's response to its Kyoto commitment. Our task was to capture value in fledgling companies by de-risking technology performance, building company capacity, and mobilizing private capital and industry partners, such that these concepts became viable products to be adopted by the market. The foundation's mandate includes technologies that contribute to cleaner air, land, and water, and, importantly, that tackle climate change. More than $3.5 billion of project funding from public and private sources managed by SDTC has developed over four hundred technology solutions (SDTC 2018). Predominantly, these solutions lie within small and medium sized enterprises (SMEs), a category of wealth generation and employment companies that contribute almost 55 percent of Canada's GDP (ISED 2019). A subset of the mature companies in SDTC's portfolio has attracted an estimated $2 billion of Follow-on Financing from the private sector. All of this can help Canada become more sustainable and competitive globally. It was a great privilege to help build these companies, working with so many great entrepreneurs across Canada and internationally.

This pragmatic mindset guides me. Solutions to environmental problems delivered by SDTC portfolio cleantech companies enable all segments of society to take action, truly integrating environmental and economic performance. Taking action builds optimism and empowerment in people to do more. It enables coalitions of the willing rather than a tragedy of the commons.

While the need for greater sustainability may be understood, the size of the opportunity is less clear. Predictions of the global low carbon market range widely. A report by the Global Commission on the Economy and the Environment in 2018 found that bold action could yield a direct economic gain of US$26 trillion from 2018 through to

2030, compared with a business-as-usual approach (Mountford et al. 2018). The reality is that what is classified as low carbon is growing and the market is dynamic. This is a critical arena for contributions from academia.

As you can see, I am an outlier among the authors in this book, albeit highly welcomed to the erudite discussion on how academia can contribute to the field of sustainability.

Overview

This chapter is written from an innovation and private sector perspective from the world outside academia. At its crux, it advocates for academic institutions to broaden their scope of study, cross-fertilize multiple disciplines, and reach out to investment, SMEs, and communities.

To assist in explaining a potential repositioning of academia's future roles, this chapter describes the innovation ecosystem writ large, along with the associated risk factors. True innovation encompasses research, development, demonstration, deployment, and diffusion into the market, which may be called RD4 rather than the more narrowly defined research and development (R&D). The chapter outlines key players on the idea creation side, complemented by the funding and private sector investment roles placed along the innovation chain. It highlights barriers and opportunities particular to sustainability. It references some innovation performance indicators for Canada and the US. The RD4 innovation ecosystem and the various actors are used as a basis to rationalize and describe specific examples of changes to academic practices. These ideas have the potential to increase economic and environmental returns to universities and colleges, SMEs, investors, and society in general. This is true innovation. The chapter also provides real-world examples of the successful evolution of academia into sustainable development markets.

The Innovation Ecosystem and Academia's Evolving Roles

While novel concepts may be great, if the consequence is not a product, process, or service that enters the marketplace, then it cannot properly be called innovation. So where do universities and colleges fit in? Is their current placement open to change? How does the critical lever of finance both intersect with universities and colleges and impact the role of academia, now and in the future?

FIGURE 6.1: Innovation Risk Profile

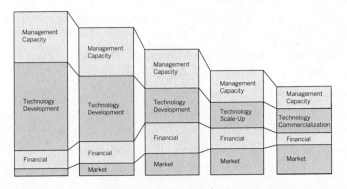

Credit: Created by Vicky J. Sharpe (adapted from "Innovation Chain" figure in SDTC 2003, 9).

This ecosystem, the result of an evolutionary and iterative process, can be shown in many ways. For simplicity's sake, it has been depicted in Figure 6.1 as having five stages in a linear format before achieving market entry, followed by increasing product sales volumes as the market adopts the innovation. From a risk perspective, many challenges need to be overcome, with their importance varying along the innovation chain. These challenges fall into four main risk categories: technology development risk, finance risk, market risk, and management capacity risk. Typically, academia is recognized as the starting point for R&D of those concepts, although industry does fund specific early stage developments that meet its needs. Academia generally focuses on reducing technology risk at the fundamental and applied research stages/prototype development. It is important to avoid the "better mousetrap" syndrome, where it is assumed that if one builds a technology well, the market will adopt it—often this is not the case. Researchers tend to pay less attention to the path that a concept will need to take to reach market entry. Market requirements include customer performance specifications, industry-wide global codes and standards, and the likelihood of meeting financing criteria imposed by the investment community.

An issue for innovation in general is that, while accurate answers to customer and market needs are not possible during the R&D phase, they are important considerations that should be addressed throughout the innovation chain once they are identified. As technology risk is diminished and management capacity strengthened, the financial and market risks become more dominant. The nature and risk appetite of investors varies across the ecosystem. In the post R&D stage, venture capital (VC) predominates. When technologies commercialize and start to produce revenues, more risk averse sources such as private equity (PE), and later on banks, step in. Greater technology/product adoption is centred on addressing market risk. If a technology is not meeting customer needs or desires at a price that they are prepared to pay, then the company will fail. Note that the use of clearly-defined boxes in the figures is for ease of illustration—the real world rarely sits neatly in boxes!

In Figure 6.2, essentially the same stages of innovation are used with technology development and demonstration (D&D), representing the stage where academia traditionally plays a small role. Demonstration can be defined as the testing of technology performance in real-world applications outside of a laboratory. The figure shows the roles of the key financing players for cleantech across the innovation ecosystem. Note that the thickness of the wedges indicates very conceptually the relative amounts of capital from the different sources.

The available funds at each stage define the "funding intensity" line. This indicates, very simplistically, the "relative," not the actual, amounts of funding. A more accurate depiction would have a much-elevated intensity line on the right side of the figure, where the stack of conventional investors/lenders provide significantly more capital when risks are substantially diminished. The shape of the funding intensity line shows an asymmetrically shaped dip or valley, which is often described as the "valley of death" for start-up companies. It occurs as conventional R&D work and funding declines and before the opportunity is

FIGURE 6.2: The Original Innovation Chain for Cleantech

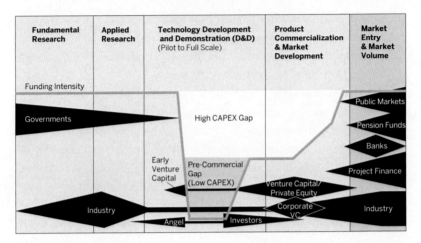

Credit: Created by Vicky J. Sharpe (adapted from "The Innovation Chain" figure in SDTC 2002, 7).

sufficiently developed for large-scale investors to become involved. Typically, this D&D phase is carried out by a small team of technology experts outside academia, funded by family and friends. Some fortunate teams can attract angel investors at this point, who bring both money and commercialization expertise. I address this later in the chapter.

Figure 6.2 illustrates that, for funding, the riskiest part of innovation is at the D&D stage, where start-ups fail, usually taking with them the value of the early stage investment. This is a market failure that requires an appropriate policy response. Yet governments, policy-makers, and academia typically continue to focus on the earlier stages. Statistics on R&D spending are proudly highlighted by governments, yet the conversion of that research into investable and eventually bankable companies does not attract the same attention. Even Business Expenditures on R&D (BERD) is more talked about than commercialization success. Data in this area is contested and infrequently collected; however, the Conference Board of Canada stated in 2018 that BERD in Canada had declined over the past fifteen years, in contrast to the growth experienced in many of its OECD counterparts, with BERD representing 0.9 percent of its GDP, versus about twice that in the United States. A small consolation is that the gross domestic expenditures on R&D (GERD) for Canada was expected to edge up 1.3 percent from the 2017 preliminary estimate to $34.5 billion in 2018 (Statistics Canada 2018). Comparing these numbers to those in the following paragraph, it can be seen that R&D spending still outstrips D&D and VC investments, which is problematic when many technologies need to scale up in size and run pre-commercial demonstrations before the capital markets will step in.

The United States is known as an innovation powerhouse. American VC investments were about US$131 billion and PE investments were US$713 billion in 2018 (NVCA 2019). The values for Canada were C$6.2 billion for VC investments and C$19 billion for PE funds in 2019 (CVCA 2020). As a simple way to make an approximate direct comparison, if one multiplies the Canadian values by ten and converts the currency, then Canada's VC and PE spending was about 35 percent and 20 percent of that of the US, respectively. This is a significant part of why commercialization success rates are lower in Canada, as converting dollars into viable companies that provide jobs and wealth is the true test. Based on these metrics, many countries are not performing well. The outputs are what matter.

Part of the reason is that, as technologies move beyond academia and are built into fledgling companies, the availability of well-accepted government support through R&D diminishes rapidly. There are a few investors, namely, "family and friends," angel investors, and a limited number of early stage (VC) funds. It is apparent that most private capital is only available when all risk elements have been largely addressed. Since the "valley of death" occurs at the D&D stages, it seems obvious that public funding is needed there to progress early companies along the innovation chain and increase the likelihood of returns on prior R&D efforts. SDTC is just such an instrument that bridges the "valley of death" by funding the "go-to-market" consortia that are addressing environmental challenges. Optimum conditions for D&D projects include a consortium of stakeholders, such as end use customers, manufacturers, and representatives of supply chains and distribution networks pertinent to specific markets and downstream investors. Early stage funds (VCs), once familiar with the small company team and the technology performance, are more likely to respond to strong results with greater investment, enabling faster market entry, i.e., the deployment stage. Governments seeking diverse and resilient economies should reduce their focus on R&D and increase direct support for D&D, reducing the imbalance between the two.

While I appreciate the view that fundamental, not industry-driven, research is valuable, and that long-term, potentially disruptive research is an important part of the ecosystem, in my view this is not a dominant part of cleantech advancement. There will be naysayers who believe that innovation should be entirely driven by the market and private sector. While it is true that some markets and technologies do not require subsidies or policy support by governments, certain ones do. Besides, has anyone found a perfect market yet? What I am espousing is a shift toward D&D, where both governments and academia can play vital roles.

A worthy role model showing the positive contribution of academia and its role in D&D to field test the performance of carbon capture and sequestration technologies occurs under the auspices of CMC Research Institutes Inc. (CMCRI), which involves multiple players. The University of Calgary provides funding, leadership, and staffing; the federal government's Western Economic Diversification Canada supplies funding; and Alberta Innovates (in the Alberta government), industry across Canada, and contributors from six countries provide expertise (CMCRI 2017). This team is developing a means of ensuring proper carbon sequestration in field operations in which oil sands companies have a big stake, and it uses the site to train students at graduate and post graduate levels.

This brings us squarely back to sustainability and low carbon economies. We are currently using the planet's resources in amounts that exceed its capacity. We must find ways to become more efficient. Cultural shifts in values and behaviours are slow, so assertive policies are needed. Currently, countries are adopting market-based instruments to drive change at varying rates, sometimes because of economic gains rather than for reasons of sustainability. In fact, these two aspects are interdependent and inextricably linked, and separating them is wrong. Canada was an early entrant when the federal government created SDTC in 2001.

The vast majority of D&D projects funded by SDTC are led by SMEs, where the intellectual property (IP) was created by entrepreneurs, many of whom had left large companies because they saw possible solutions to industry problems and wished to work for themselves. This may be a defining characteristic of cleantech companies. In the first decade of SDTC operations, my calculations showed that only 12 percent of background IP came from universities. However, academia contributed in conventional ways to SDTC projects by providing expert advice and use of facilities. For example, the University of New Brunswick was part of a seven-member consortium determining the efficacy of using hydrogen-enriched natural gas in internal combustion engines for power generation. Cleantech is a growth sector, so academia needs to look at how it can make a greater contribution by linking up with start-up companies and angel investors within its local communities.

While many companies face this challenge (other than software-based companies, which have small capital equipment requirements, or "low capex"), innovative sustainable technologies, or "cleantech," face another funding gap. Cleantech technologies are often energy or water related and thus require significant investment in capital equipment, hence the label high capital equipment technology, or "high capex." Few VCs are able to provide sufficient capital and wait out longer development/investment cycles. Additionally, the technology must be economically viable at full scale, meaning that several steps to scale up the technology are required before the product is at commercial scale or production levels. The investment community's general risk aversion and its preference for natural resource investments in Canada have created a "high capex" financing gap for cleantech/energy technologies (see Figure 6.2).

If you are wondering why you should care, it is these high capex technologies that often have greater greenhouse gas (GHG) reduction potential. Also, because energy and water technologies are "infrastructure like," they provide more durable and lasting benefits that should resonate with investors who are looking for predictable long-term returns that should include the likes of pension funds. These institutional investors have been slow to integrate sustainability as an opportunity for better returns and a risk management tool into their investment screening criteria (Sharpe 2015), although the tide is turning. The Canadian government created the Canada Infrastructure Bank in June 2017, with C$35 billion to attract and co-invest with the private sector and institutional investors in new, revenue-generating infrastructure projects. In April 2020, the Canada Infrastructure Bank (2020) announced its participation in nine projects in sustainable transit, renewable energy, and clean water and waste water, and it had committed C$3.6 billion to four of these projects, which should improve this situation for sustainable technologies.

While academia does not traditionally play at this stage of the ecosystem, understanding these growth capital challenges and helping to strengthen certain aspects of the value proposition early on are needed. Testing facilities for larger-scale technologies can be provided by appropriately designed laboratories that can both mimic field conditions and be customizable for different applications. A positive example backed by the federal government is the Canada Foundation for Innovation (CFI 2020), which invests in research facilities and equipment in Canada's universities, colleges, research hospitals, and non-profit research institutions. A private sector example is that of the Verschuren Centre for Sustainability in Energy and the Environment, which sprang from a close relationship with Cape Breton University. It works with Indigenous and other local communities, academics, SMEs, and global companies on areas such as carbon materials transformation and the bio-economy, using equipment platforms for testing and commercialization activities.

The business case for clean/green technologies is continually strengthening, causing governments globally to implement policies addressing climate change, often focusing on either "scale-up" capital for technology push or consumer-oriented incentives to create market pull. This is resulting in more funding across the innovation chain. Consequently, the need for more people entering finance and policy roles is increasing, and academia can help build cohorts of highly qualified personnel with the necessary skill sets to fulfill these roles.

There are many important ways that academia can contribute to more effective innovation. The following examples demonstrate how universities and colleges can move closer to market on the innovation continuum, encompassing the view that we should care about RD4, not just R&D. Furthermore, contributions are required from more than just the obvious disciplines, such as biology, ecology, environmental sciences, engineering, medicine, and natural resource management. Contributions from the social sciences, arts, finance, communications, political science, and many more are important ingredients for success. The suggestions have been grouped into the following areas:

- build multidisciplinary, cross-departmental research teams and curricula

- develop and deliver entrepreneur training programs
- change reward systems for professors
- change IP development and ownership systems
- increase interaction with entrepreneurs
- integrate policy development, life cycle analysis, and performance evaluation into curricula
- increase interaction with the local community, investors, and SMES

Build Multidisciplinary, Cross-Departmental Research Teams and Curricula

A sustainable technology is superior if it provides holistic solutions, which, by definition, often requires a multidisciplinary approach to problem solving. Typically, professors compete for grants in specific disciplines, and this can discourage cross-disciplinary innovation research. While there has been a greater emphasis on harnessing skills across disciplines and increasing the numbers of cross-appointed professors, more is needed at university and funding agency levels. Canada's New Frontiers in Research Fund and the Advanced Research Projects Agency-Energy (ARPA-E), a US government agency tasked with promoting and funding research and development, are moves in the right direction.

Develop and Deliver Entrepreneur Training Programs

The much-debated topic of whether an entrepreneur is born or trained seems unnecessary, because even if some students have a predilection for risk, they can always manage that process better when taught. Many companies start out being driven by technologists; however, to succeed, they need complementary skills of human resource management, financing, marketing, communications, and business development strategy, to name a few. Building diverse skill set teams, as well as plain old diversity, is known to improve outcomes, yet this does not happen enough within academia. While it can be argued that academia is not in the business of building companies, that does not resonate when we realize how much we need better, sustainable, cost-effective solutions to our current problems.

A prime example of integrating academics into company building and improving performance is the Creative Destruction Laboratory

(CDL). This was started in 2012 at the Rotman School of Management, University of Toronto, and is now offered at universities across Canada, plus locations in the UK, the US, and France. The CDL program pairs start-up founders with experienced technology entrepreneurs and investors, both institutional and private. It focuses on designing a series of company objectives, such as building teams or raising capital, that have measurable deliverables. Accountants and lawyers also lend support. Graduating companies become more mature in their approach to markets and frequently see increased investment.

Change Reward Systems for Professors

Human resource performance metrics and rewards drive behavioural change. Businesses often recognize a dual ladder where different contributions are equally recognized, such as technical versus management. Some professors are superb teachers, a difficult skill, and others have stronger research skills. Why not let academics and their students become wealthy as a result of a successful company spin-out? Let them follow the company and be away from core teaching for a while, returning to inject knowledge, and possibly wealth, into the university system. This approach occurs far more frequently in the US than in Canada, with University of California, Los Angeles (UCLA) being a good case in point. The golden horseshoe around Toronto, running from Waterloo to Kingston, is also making great strides in developing clusters of entrepreneurs with proper support systems, with academia playing a significant role.

Change IP Development and Ownership Systems

Academia tends to have a narrow view of technology development and IP creation. Counting the number of peer reviewed, published papers is important in grant applications, yet this may be the worst way to protect IP and stress test an emerging company for IP protection. Applied research, with industry participation and funding, still generates excellence; it is just that the selection and recognition process should be different. Not all top performing professors or students need to publish. This culture shift, likely driven by universities, will need to be accepted by the granting councils.

Another challenge is the way academia handles IP ownership and payments for providing facilities. Universities and colleges have encumbered IP creation, registration, and protection by having many different ways to derive payment/ownership of the IP. It is a well-known frustration among venture capitalists that taking IP out of academia is often a lengthy and expensive proposition. As time to market is important for investors and inventors alike, this issue is important and far from ideally handled. Standardization of IP registration and handling across academia would be really helpful.

Training entrepreneurs within and external to academia to understand what IP to develop and how to register and protect IP is critical, for without this the companies have limited value. Again, there are a plethora of organizations acting as "incubators" and "accelerators," and they would benefit if universities and colleges integrated efforts to support young companies to grow, test, and protect their IP. Having a critical mass of start-ups in a single location enables sharing ideas and aggregation of value propositions. At their best, these shared strategies on how to get to market and serve customers will strengthen multiple SMEs acting in concert. Another opportunity is to develop shared IP strategies so that a group can influence the creation of specific standards or regulations that specify their new IP, helping to lock in markets. Participating on standard setting committees and helping to set pricing levels for licensing of shared IP are means of creating a wedge into the market to oust competition or incumbent products. Generally, large institutions and governments set the frameworks, but aggregated start-ups can provide the content and direction that serve their best interests by influencing the standard and regulatory setting processes. Academic institutions could become more involved in these arenas, in concert with SMEs, rather than sitting as individual experts on such bodies. Academia could become a platform for greater IP success.

Increase Interaction with Entrepreneurs

Universities in the US and Canada have already made great strides in strengthening commercialization success, as evidenced by year over year increases in patents applied for, patents awarded, issuance of licenses and associated revenues, number of start-ups, and product sales. As a rule of thumb, Canada is performing as well as the US (divide the numbers in the US by ten), as measured by survey data collected by the admirable AUTM (formerly called the Association of University

Technology Managers). In 2017, for example, 34 Canadian institutions reported a record high of 111 new start-ups, and 193 US institutions reported 1080 start-ups (survey response rates were 49 percent and 62 percent for Canada and the US, respectively) (AUTM 2017). However, much greater success is needed for these systems to become more self-sustaining. Competition for government largesse is increasing, and R&D is not winning this race. The licensing revenues relative to R&D funds expended are still poor in Canada. Academic institutions in the US perform well at this, where individual players like the Massachusetts Institute of Technology alone deliver returns at a greater level than those from an aggregation of Canadian universities. Bringing in senior serial entrepreneurs to universities and colleges to interact with researchers and fledgling companies adds tremendous value for everyone.

Integrate Policy Development, Life Cycle Analysis, and Performance Evaluation into Curricula

Governments at all levels have a huge impact on markets, whether through regulation, standards, laws, program design, urban design, or just plain politicking. Like many institutions, they are risk averse and yet are designed to absorb risk. This presents an interesting conundrum. Regardless, it is a common refrain that policies appear to be designed in a commercial and economic vacuum. Universities could make a significant contribution to broadening knowledge and train students to better appreciate the application of policy and its market impacts. Full life cycle analysis to support decision making is known to be a superior approach, yet there are few practitioners. Consumer decisions about which technologies to purchase are made difficult because appropriate information is not available to the public. Regulatory systems do not encourage innovation because of narrow and fragmented thinking. Academia could help to champion these attitudinal changes and then build skills in the student body to deliver better policy. Yes, there are famous policy institutes, yet dare it be said that some are too academic!

Governance principles for corporations, governments, not-for-profit entities, and SMEs require upgrading to incorporate sustainability. International and local trade agreements and practices likewise need better integration of sustainability principles. Destruction of habitats, impoverishment of sections of society, and corruption continue unabated because we do not have appropriate guidelines or enforcement. Again, application of cross-faculty skills and research, interfaced with non-academic players, would benefit these important and complex arenas.

A positive example of academia's influence in the policy sphere is an initiative called Smart Prosperity. This group of influential people drawn from industry, the financial sector, innovators, and ex-politicians and bureaucrats is fully supported by the Smart Prosperity Institute, which is housed within the University of Ottawa. The institute's staff undertake the background research that enables the outward facing group to take logical, fact-based positions on climate change and the innovation of government programs. I am involved in content on innovation and investment ecosystems and have observed firsthand the thoughtful and powerful influence that would not have been possible without its academic core team.

The entire science, and I would argue art, of evaluating the performance of government programs to strengthen the economy and/or sustainability of our society does not have enough quality research or data. International performance benchmarks of any veracity are hard to come by. Data collection is inconsistent, and baselines for determining results are rarely properly collected. The track record in the US and Canada suggests that industry is good at this for more straightforward product performance, and yet for highly influential, billion-dollar spending/investment programs, the record is not strong. Again, academia can help by bringing multiple disciplines and industry/industry associations together to build repositories of baseline and program impact data.

Increase Interaction with the Local Community, Investors, and SMEs
Cultural shifts for greater sustainability are required across societies. Young adults want society to be more responsive to issues of equity, diversity, and sustainability, and they want to work with and for entities that have these characteristics. Academia can provide curricula and community engagement that help to answer these needs. While universities and colleges reach out to their communities, it is often for fundraising; this new outreach is intended to be different. Universities are communities in themselves and can often be isolated on campuses;

greater interaction with the surrounding urban centre would likely benefit everyone. Building start-ups locally will attract investments and create local high-quality jobs, further demonstrating the value to the community of having a university or college in their midst.

Social inequality, and how governments, industry, civil society, and universities should address these often-divisive issues, requires sensitivity to and the construction of real relationships with diverse and less privileged communities. An example of this is an attempt to raise funds for an Indigenous infrastructure fund where academia can contribute expertise and objective perspectives to problem solving, especially where there are local urban Indigenous Peoples in the university/college catchment area (Isaac and Sharpe 2020). Furthermore, distance learning and skills development for more remote locations is another area where academia is highly relevant.

Capital markets, as efficient and selfish as they are, have been slow to integrate sustainability into investment criteria. This has left portfolios of assets with a high carbon content, for example, fossil fuel extraction and end use, that are becoming costly stranded assets as these industries are disintermediated, dragging down investment returns. Leading entities have recognized this, and in response to the recommendations of the Task Force on Climate-related Financial Disclosures (TCFD), "nearly 800 public- and private-sector organizations have announced their support for the TCFD and its work, including global financial firms responsible for assets in excess of $118 trillion" (TCFD 2019, i). Hence, there is a significant increase in the demand for expertise to respond to new requirements. Regulators are seeking change, and companies need to understand the implications. Again, academia can assist with this in integrating sustainability into many disciplines, such as economics and MBA curricula. Also, they can upgrade their pension portfolios to reduce climate risk.

Co-op programs are an excellent means of placing students at the coal face of creativity, particularly if they work inside SMEs, not large corporations. While persuading enough SMEs to take students may be difficult from an affordability perspective, start-ups seek skilled help and young minds. While the compensation may be small, the lessons learned are valuable. Additionally, it could be a productive mix for academia to provide shared space for students, service providers, and start-ups. A superior example is Ryerson University's DMZ (formerly Digital Media Zone), an accelerator for tech start-ups that was ranked by UBI Global as one of the top five university business incubators in the world (Meyer and Sowah 2019). Another example is Communitech, an industry-led incubator with strong ties to the University of Waterloo.

Conclusion

I have described where academia currently sits in the innovation ecosystem and how it might change to address more areas. I believe universities can foster innovation in a manner that will attract more financing and see greater returns, human and financial, for educational institutions and society at large. I have highlighted investable opportunities where academia can continue to add knowledge, create multidisciplinary results, and help support entrepreneurs to deliver solutions to market, i.e., true innovation, for a more sustainable world. Hopefully these ideas are helpful grist for the academic mill.

References

AUTM. 2017. *2017 Licensing Activity Survey*. https://autm.net/surveys-and-tools/surveys/licensing-survey/2017-licensing-activity-survey.

Canada Infrastructure Bank. 2020. *Q4 2019–20 Progress Update*. April 2020. Accessed May 10, 2021. https://cib-bic.ca/wp-content/uploads/2020/06/Q4-CIB-placemat-Final-EN.pdf.

CFI (Canada Foundation for Innovation). 2020. "Funded Projects." Updated August 2020. https://www.innovation.ca/funded-projects.

CMCRI (CMC Research Institutes). 2017. *CMC Research Institutes Annual Report 2016–2017*. Published October 2, 2017. https://issuu.com/cmcghg/docs/2016-2017_annual_report_final_issuu.

Conference Board of Canada, The. 2018. "Business Enterprise R&D." Updated May 2018. https://www.conferenceboard.ca/hcp/provincial/innovation/berd.aspx.

CVCA (Canadian Venture Capital and Private Equity Association). 2020. *Year End—2019: Canadian VC & PE Market Overview*. March 11, 2020. https://www.cvca.ca/research-insight/market-reports/.

Isaac, David, and Vicky Sharpe. 2020. "An Indigenous Infrastructure Fund Can Offer Lasting, Meaningful Change in Canada." *The Globe and Mail*, September 14, 2020. https://www.theglobeandmail.com/business/commentary/article-an-indigenous-infrastructure-fund-can-offer-lasting-meaningful-change/.

ISED (Innovation, Science and Economic Development Canada). 2019. *Key Small Business Statistics—November 2019*. Ottawa: ISED. https://www.ic.gc.ca/eic/site/061.nsf/vwapj/KSBS_Nov-2019_En_Final_5.pdf/$file/KSBS_Nov-2019_En_Final_5.pdf.

Meyer, Holger, and Joshua Sowah. 2019. *World Rankings 19/20 Report*. Stockholm: UBI Global. Published November 2019. https://resources.ubi-global.com/hubfs/Publications/Rankings/UBI%20Global%20-%20Rankings%201920%20v2.pdf.

Mountford, Helen, Jan Corfee-Morlot, Molly McGregor, Ferzina Banaji, Amar Bhattacharya, Jessica Brand, Sarah Colenbrander, Ed Davey, Laëtitia de Villepin, Faustine Delasalle, Annabel Farr, Leonardo Garrido, Ipek Gençsü, Saira George, Catlyne Haddaoui, Leah Lazer, Nathaniel Mason, Jeremy Oppenheim, Rachel Spiegel, Lord Nicholas Stern, and Michael Westphal. 2018. *Unlocking the Inclusive Growth Story of the 21st Century: Accelerating Climate Action in Urgent Times*. Published August 2018. Washington, DC: New Climate Economy. https://newclimateeconomy.report/2018/wp-content/uploads/sites/6/2019/04/NCE_2018Report_Full_FINAL.pdf.

NVCA (National Venture Capital Association). 2019. *National Venture Capital Association 2019 Yearbook*. Data provided by PitchBook. https://nvca.org/wp-content/uploads/2019/08/NVCA-2019-Yearbook.pdf.

SDTC (Sustainable Development Technology Canada). 2002. *Partnering for Real Results: 2002 Annual Report*.

SDTC (Sustainable Development Technology Canada). 2003. *Initial Plan: April 2003*.

SDTC (Sustainable Development Technology Canada). 2018. *Accelerate Canadian Cleantech Companies on a Faster Path from Startup to Scale Up—Annual Report 2017–2018*. https://www.sdtc.ca/wp-content/uploads/2020/01/ASCRIBE_18-207_AR_e_WEB.pdf.

Sharpe, Vicky. 2015. "Protect Your Pension and the Planet." *Corporate Knights* (Winter), January 19, 2015. https://www.corporateknights.com/responsible-investing/protect-pension-planet/.

Statistics Canada. 2018. "Spending on Research and Development, 2018 Intentions." *The Daily*, December 12, 2018. https://www150.statcan.gc.ca/n1/daily-quotidien/181212/dq181212c-eng.htm (archived content).

TCFD (Task Force on Climate-related Financial Disclosures). 2019. *2019 Status Report: Task Force on Climate-related Financial Disclosures: Status Report*. June 2019. https://www.fsb-tcfd.org/publications/tcfd-2019-status-report/.

WCED (World Commission on Environment and Development). 1987. *Our Common Future*. Oxford, UK: Oxford University Press.

7
Sustainability and Decision

THOMAS DIETZ

WHAT DO WE MEAN BY SUSTAINABILITY? At the most generic level, the idea of using natural resources so they will not be depleted stretches back to treatises on forestry and mining in the seventeenth and eighteenth centuries.[1] The 1916 law founding the US National Park Service may have been the first North American legislation asserting that present use must be balanced with the needs of future generations. In the 1960s, inspired in part by Rachel Carson's *Silent Spring* (1962), the environmental movement became broader and more engaged with environmental justice and moved toward the ideas that are the basis for sustainability (Dietz 2020). In 1980, a major report from the International Union for Conservation of Nature (IUCN 1980), led by Lee Talbot, called for the integration of conservation and development based on science. Then came the Brundtland report in 1987 and the notion of sustainable development (WCED 1987). Concerns with the commitment of sustainable development to growth led some to draw a distinction between sustainability and sustainable development, although often the terms are used interchangeably (Robinson 2004).

There are thousands of definitions of sustainability. Following the logic of Dewey, Pearce, and the American pragmatists, I want to ask what "work" an idea of sustainability can do for us (Misak 2013). In particular, I want to ask how any particular view of sustainability will facilitate decision making.² Decisions that involve human well-being, the well-being of other species, the state of the environment, the needs of future generations, and all the other issues that swirl around sustainability involve complicated trade-offs. A useful concept of sustainability will help us think through those tough decisions.

My sense is that there are five common definitions in use:

- *Do good.* Sustainability in some ways has come to mean all things to all people, and in that sense it is little more than a vague admonition. Needless to say, this approach provides little guidance—more specificity is needed.
- *The Brundtland definition.* Sustainable development is a process that "meets the needs of current generations without compromising the ability of future generations to meet their own needs" (WCED 1987, 23). The issue arises as to how to implement this in development projects. Another issue is that it leads to the multiple capitals approach.
- *Multiple capitals and weak versus strong sustainability.* This analytic framework posits that human well-being is produced by deploying manufactured, human, natural, and social resources.³ The centre of this literature is about strong versus weak sustainability and whether or not these resources are substitutable across categories, and, even if they are, whether it is ethical to allow some kinds of resources to be depleted (Neumayer 2010). These kinds of analyses get very technical very quickly, so a simpler approach emerged in the triple bottom line.
- *The triple bottom line.* Starting with Elkington (1997, 2004), the idea is that business and governments should make decisions not just to maximize profits or grow the economy, but should also take into account the impacts of decisions on people and the environment. The problem is that it is not clear how trade-offs across the three bottom lines should be made. For example, if a company can improve its environmental performance, but only at the cost of eliminating jobs critical to a local community, how does one balance these two "bottom lines" in making a decision?

- *Environmentally efficient well-being.* This view suggests that we have two primary ethical goals: increasing human well-being and decreasing harm to the environment (Dietz, Rosa, and York 2009; Jorgenson 2014; Steinberger et al. 2012; Jorgenson and Dietz 2015; Roberts et al. 2020). This literature acknowledges that the environment and human well-being are both multidimensional, and that human well-being is not the same as affluence. Part of this approach is the growing concern that gross domestic product per capita or similar measures of economic scale are not adequate to capture well-being. It follows that decisions should take account of both criteria, try to find optimal regions in the resulting two-dimensional space (human well-being vs. environment), and be explicit about trade-offs.

  Environmentally efficient well-being is my preferred conceptualization of sustainability. But my intent in raising it and the other four is to highlight the point that sustainability is about decision making. Thus, I will suggest strategies that emphasize that decision making is the core of sustainability, and that sustainability science is intended to support decision making.

I came to the university as a first-generation white male college student from a working-class family, with only a very vague idea of how knowledge was or could be organized. As an undergraduate at Kent State, I became committed to human ecology as a means of drawing together scholarship across disciplines to understand humans in ecosystems and the biosphere. I was able to help design and implement an undergraduate led study of the Cuyahoga River that integrated work in hydrology, limnology, environmental toxicology, sociology, and economics (Barone et al. 1971). That cemented my commitment to interdisciplinary work. I went on to pursue a PHD in ecology, while identifying with the then emerging field of environmental sociology. Nearly all of my academic career has been at public universities, where I have always had appointments in both interdisciplinary environmental programs and sociology and have been active in creating several new interdisciplinary programs. But my perspectives were also shaped by organizing in the interactions among the evolving environmental, anti-war, free university, civil rights, and other social movements on college campuses. I was deeply involved in the turbulent politics before and after the shootings at Kent State in the spring of 1970, less than a month after the first Earth Day (Dietz 2020). The challenges of those times made it clear that uncertainty is central to human life and must be considered in all our analyses. Those experiences also gave me a vision of the university as a place that engages society around both facts and values and that should pioneer effective ways of deliberating about both.

Challenges to Sustainability Science Research and Education

Sustainability science research is intended to both advance fundamental knowledge and be of practical value (Kates et al. 2001; Stokes 1997). What are the challenges to advancing sustainability science? I believe the same challenges face both research and education. Indeed, as a first principle, we should be linking research and education as closely as we can manage.

How will we know if we have been successful in meeting these challenges? I would suggest that a decade or two hence we would observe that major societal decisions are well informed by scientific knowledge. We also would observe that attention is given to differences in values, to uncertainty in facts and values, to fairness in process and outcomes, and, in particular, to the effects of our decisions on those who are most vulnerable (Dietz 2013a, 2013b). Processes for decision making would engage with science and with interested and affected parties. Our universities would provide a knowledge base and tools for decision making, and they would also have educated large numbers of leaders in government, the private sector, and civil society who can deploy those tools. But to get there, we have to overcome several challenges.

Interdisciplinarity

We have to foster interdisciplinarity. For fifty years, we have known that interdisciplinarity is hindered by university, government, and professional organization structures that allocate budget, prestige, and professional identity primarily to disciplinary units (McEvoy III 1972). There are many ways universities have tried to overcome this challenge. Some have instituted departments or colleges of sustainability.[4]

Others have tried to make sustainability a theme that permeates across the whole institution. Still others have tried to use a network structure that links and supports interested faculty without creating an organizational unit like a college or department, but also without trying to engage everyone. And, of course, there are hybrids. Certainly, what works well and what does not will depend a great deal on context. We should discipline our conjectures and anecdotes about what works and what doesn't with analyses of the natural experiment underway where universities try different approaches. There is some tradition of research on interdisciplinarity that can serve as a starting point. But we need to clarify what we mean by success and try to learn to assess success systematically.[5]

Incorporating Values

We have to deal more effectively with values. Many scientists have a tendency to think that if the public or government decision makers or corporate executives simply knew what the scientists knew, decisions would be different.[6] This leads to a call for more science communication, where too often that means getting others to agree with the scientists. But decisions are almost never made on the basis of facts alone. They always involve values (Dietz 2013a). I may agree with you completely on the facts but support a different course of action because my values are different. Further, my values and my experience may make me concerned with a different set of factual questions than the ones you have investigated. You may have the science right, but you might not have the right science to inform my decision. This implies we need to be discussing, and studying, both facts and values and learning about the values of concern to those making decisions and for those interested in or affected by a decision.[7]

Coping with Uncertainty

Applying science to decisions nearly always increases uncertainty (Rosa 1998; Rosa, Renn, and McCright 2013; Dietz 2013b; York 2013). The toxicity of a compound may be well understood in the laboratory, and global mean surface temperatures may be projected with reasonable certainty. But decisions have to be made about the risks of toxic exposure via complex and not well-documented pathways in a community already exposed to other health stressors. Climate adaptation depends on very local trajectories of temperature, patterns of precipitation, and much else. Moving from the global to the national, regional, and local scales where decisions are made increases uncertainty. That requires a dialogue that allows scientists to understand the local context and those who have to make decisions to understand the science and its strengths and limits. Those interested in and affected by a decision, the public, have to be part of the discussion (Dewey 1927). The dialogue builds understanding and trust (National Research Council 2008).

Linking Analysis and Deliberation

These last two problems suggest that we have to embrace research processes that link scientific analysis to deliberation with decision makers and interested and affected parties, what is sometimes referred to by the shorthand phrase "analytic deliberative process" (Dietz and Stern 2008; Dietz 2013a; National Research Council 2008). Linking research to public deliberation has several advantages.

First, since decisions are never made on the basis of facts alone, but always involve values, we need a broad assessment and discussion of values that interested and affected parties bring to the decision and mechanisms that might facilitate decision making in the face of value differences. Just as we need to accept and deal effectively with scientific uncertainty, we have to accept and deal effectively with value uncertainty. How to do this can be a research topic itself.

Second, since most sustainability decisions are made in very specific contexts, scientific understanding has to be adapted to that context. This requires detailed local knowledge and a shared sense of the increased uncertainty that comes from taking general scientific understanding and applying it to a specific context. Thus, engagement with those who are experts on local contexts, including traditional ecological knowledge, helps get the science right (Whyte 2013; Whyte, Brewer II, and Johnson 2016). The logic of standpoint theory can help us think through these matters (Harding 2015).

Third, a process of ongoing deliberation between researchers and interested and affected parties can help ensure that the science being done answers questions important to decision making. It helps get the right science.

Finally, the ongoing interaction among researchers, decision makers, and interested and affected parties helps build mutual understanding and trust. That, in turn, makes it more likely that research will influence decision making. Over time, overall capacity for decision making improves.

This challenge needs to be met institutionally. We should not expect every researcher to spend substantial amounts of time in deliberation with the public and decision makers any more than we should expect everyone to become skilled bloggers or public debaters. We have some models for ways to link science to decision making in the form of the land-grant institutions, the NOAA funded Regional Integrated Sciences and Assessments climate adaptation centres,[8] and other efforts to link research to decision makers and the public. Again, we have to assess what works and what doesn't and how that varies across context.

As we develop sustainability as a basis for both decision making and programmatic efforts at universities, we have to avoid becoming inward looking and isolated. Speculation about the future is always both necessary and error prone. But it seems plausible to me that over the twenty-first century some of the biggest influences on both human well-being and the environment will come from technological change. We are experiencing a revolution in artificial intelligence, biotechnology, nanotechnology, cognitive science, and robotics. These have the potential to transform human life and to have massive effects on the environment. They may provide solutions to some of the most vexing problems of sustainability. They may also create problems at least as challenging as those we face now.

Dealing with sustainability gives universities and society a testbed in which to develop approaches for dealing with these emerging transformations. If we can find ways in research, education, and engagement to adequately foster interdisciplinarity, engage values, and cope with uncertainty, we may provide guidance as to how to structure our work around the emerging issues. And, at a minimum, we should engage with these emerging transformational areas of science so that we can learn from them and they from us.

Conclusion

In our scholarship and teaching, we want to advance fundamental understanding. But we also want to contribute to improved societal decision making. Donald Stokes (1997) pointed out that many see a conflict between contributing to fundamental understanding and being useful in examining the goals of research. His analysis, with some modifications I have made, is represented in Figure 7.1.

FIGURE 7.1: Pasteur/Ostrom's Quadrant

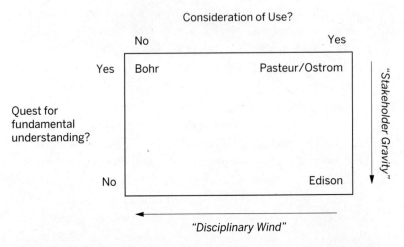

Credit: Created by Thomas Dietz (adapted from Figure 3-5, Quadrant Model of Scientific Research, in Stokes 1997, 73).

Stokes noted that some researchers care only about contributing to fundamental knowledge, and he labeled that corner Bohr's Quadrant after Danish physicist and quantum theory pioneer Niels Bohr (Ottaviani and Purvis 2009). Some care only about being useful and not about fundamental knowledge, and that quadrant Stokes named after Edison, who pioneered the applied research laboratory. But he noted that one can do both, and that approach he labeled Pasteur's Quadrant, since Louis Pasteur made foundational contributions to microbiology while also finding ways to vaccinate against anthrax and rabies, preserve food, and improve the quality of wine and cheese.

I have modified Stokes' graphic to include Elinor Ostrom, the first woman to win the Nobel Prize for economics. Ostrom was a pioneer in understanding how commons (e.g., forests, fisheries, groundwater, and the climate) can be managed sustainably (Ostrom 2009, 2010; Dietz, Ostrom, and Stern 2003). Her work made fundamental contributions, but at the same time was always grounded in an appreciation that commons management had huge impacts on the lives of people around the world.

Why isn't it easy to occupy the Pasteur/Ostrom quadrant? I have suggested two forces that pull us away. One is disciplinary wind. The priorities that emerge from disciplines often push us away from work that is useful. I label this force "wind" because it is not constant over time but changes from time to time as new topics become popular within a discipline and other topics fall out of favour. The other pressure is stakeholder gravity. To the extent that our research and teaching agendas are influenced by the needs of those who want to use our research or hire our students, there is a pull toward what is useful. That can move us away from contributing to the fundamental knowledge we need to keep our understanding moving forward. I think this force is more constant in direction than the shifting priorities of disciplines, so I label it "gravity."

If we can be interdisciplinary, incorporate values into our work, cope with uncertainty, link analysis with deliberation, and avoid creating our own intellectual silos, we have a hope of occupying the Pasteur/Ostrom quadrant in our teaching and research. Accomplishing all that will be a challenge. It will have to be thought through in each context where we work, but that also means we can learn from each other as we go. Experimenting and learning from our experiments moves us toward social learning for sustainability, a view that we don't know what to do but that we can learn (Henry 2009). And that will poise us to deal with not only the challenges of sustainability as we see them today, but also the challenges that will emerge over the rest of the century.

Notes

1. Essays by Robinson (2004) and by Mitcham (1995) are insightful starting points for the history of the concept of sustainability, as are books by Caradonna (2014) and Grober (2012). Perhaps the most useful narratives in many ways are Worster's (2016) history of concepts of growth and limits and Macekura's (2015) history of sustainable development. Longo et al. (2016) offer a very useful analysis of the concept of sustainability in relation

to "taken-for-granted" assumptions about growth and suggest some alternative framings.
2. I am also interested in how theories of decision making inform our understanding of sustainability, and the strengths and limits of particular theories, such as the rational actor model, values-beliefs-norms theory, and the heuristics and biases framework (Dietz 1994), but that is too much to get into here.
3. While the tradition is to call these "capital," I have argued that there are analytical advantages to calling them resources (Dietz 2015).
4. These have various names, and not all use the term "sustainability." Many but not all interdisciplinary environmental programs can be thought of as first-generation sustainability programs, and much can be learned from them.
5. The simplest lesson is perhaps that skilled leaders for the program and strong support from administrators who control resources are key. Conversely, my observation is that administrators and faculty skeptical of innovation and interdisciplinarity can cause great damage. But we need more systematic assessment of how organizational forms provide the incentives and the resources for success.
6. I am reminded of Strother Martin's character, the captain, in the movie *Cool Hand Luke*, admonishing Luke to "get your mind right" and noting "What we've got here is failure to communicate" (Burbank, CA: Warner Bros.-Seven Arts, 1967). For a recent summary of research on science communication, see *Communicating Science Effectively: A Research Agenda* (National Academies of Sciences, Engineering, and Medicine 2017).
7. I would conjecture that many who are very concerned with various sustainability issues, and who have deep value commitments, may have trouble articulating the kind of ethical reasoning they feel should be used in decision making. For example, many working on the environment raise objections to the utilitarianism that underpins welfare economics, but I'm often not sure if they are arguing instead for a Kantian ethics, or deliberative ethics, a capabilities approach, or something else.
8. See Climate Program Office, "About the Regional Integrated Sciences and Assessments Program," accessed March 9, 2022, https://cpo.noaa.gov/Meet-the-Divisions/Climate-and-Societal-Interactions/RISA/About-RISA.

References

Barone, John, Robert Buller, Thomas Carrothers, Thomas Dietz, Diana Franchi, Mark Galizio, Johanna Krehbel, Richard Lytle, Ken McCluggage, Barbara Taylor, Dave Tompkins, Gary Walter, Larry Webb, and Phyllis Wolfe. 1971. *An Interdisciplinary Study of Some Environmental, Social and Economic Factors Affecting a Section of the Cuyahoga River Watershed*. Kent, Ohio: Center for Urban Regionalism, Kent State University.

Caradonna, Jeremy L. 2014. *Sustainability: A History*. Oxford: Oxford University Press.

Carson, Rachel. 1962. *Silent Spring*. New York: Houghton Mifflin Harcourt.

Dewey, John. 1927. *The Public and Its Problems*. New York: Henry Holt.

Dietz, Thomas. 1994. "'What Should We Do?' Human Ecology and Collective Decision Making." *Human Ecology Review* 1 (2): 301–9.

Dietz, Thomas. 2013a. "Bringing Values and Deliberation to Science Communication." *Proceedings of the National Academy of Sciences* 110 (Supplement 3): 14081–87.

Dietz, Thomas. 2013b. "Epistemology, Ontology, and the Practice of Structural Human Ecology." In *Structural Human Ecology: Essays in Risk, Energy, and Sustainability*, edited by Thomas Dietz and Andrew K. Jorgenson, 31–52. Pullman, WA: WSU Press.

Dietz, Thomas. 2015. "Prolegomenon a Structural Human Ecology of Human Well-Being." *Sociology of Development* 1 (1): 123–48.

Dietz, Thomas. 2020. "Earth Day: 50 Years of Continuity and Change in Environmentalism." *One Earth* 2 (4): 306–08. https://doi.org/10.1016/j.oneear.2020.04.003.

Dietz, Thomas, Elinor Ostrom, and Paul C. Stern. 2003. "The Struggle to Govern the Commons." *Science* 302 (5652): 1907–12.

Dietz, Thomas, Eugene A. Rosa, and Richard York. 2009. "Environmentally Efficient Well-Being: Rethinking Sustainability as the Relationship between Human Well-Being and Environmental Impacts." *Human Ecology Review* 16 (1): 114–23.

Dietz, Thomas, and Paul C. Stern, eds. 2008. *Public Participation in Environmental Assessment and Decision Making*. Washington, DC: The National Academies Press.

Elkington, John. 1997. *Cannibals with Forks: The Triple Bottom Line of 21st Century Business*. Gabriola Island, BC: New Society Publishers.

Elkington, John. 2004. "Enter the Triple Bottom Line." In *The Triple Bottom Line, Does It All Add Up? Assessing the Sustainability of Business and CSR*, edited by Adrian Henriques and Julie Richardson, 1–16. London: Earthscan.

Grober, Ulrich. 2012. *Sustainability: A Cultural History*. Translated by Ray Cunningham. Totnes, Devon, UK: Green Books.

Harding, Sandra. 2015. *Objectivity and Diversity: Another Logic of Scientific Research*. Chicago: The University of Chicago Press.

Henry, Adam Douglas. 2009. "The Challenge of Learning for Sustainability: A Prolegomenon to Theory." *Human Ecology Review* 16 (2): 131–40.

IUCN (International Union for Conservation of Nature). 1980. *World Conservation Strategy: Living Resource Conservation for Sustainable Development*. Gland, Switzerland: International Union for Conservation of Nature.

Jorgenson, Andrew K. 2014. "Economic Development and the Carbon Intensity of Human Well-Being." *Nature Climate Change* 4 (March): 186–89.

Jorgenson, Andrew K., and Thomas Dietz. 2015. "Economic Growth Does Not Reduce the Ecological Intensity of Human Well-Being." *Sustainability Science* 10 (1): 149–56. https://doi.org/10.1007/s11625-014-0264-6.

Kates, Robert W., William C. Clark, Robert Corell, J. Michael Hall, Carlo C. Jaeger, Ian Lowe, James J. McCarthy, Hans Joachim Schellnhuber, Bert Bolin, Nancy M. Dickson, Sylvie Faucheux, Gilberto C. Gallopin, Arnulf Grübler, Brian Huntley, Jill Jäger, Narpat S. Jodha, Roger E. Kasperson, Akin Mabogunje, Pamela Matson, Harold Mooney, Berrien Moore III, Timothy O'Riordan, and Uno Svedin. 2001. "Sustainability Science." *Science* 292 (5517): 641–42.

Longo, Stefano B., Brett Clark, Thomas E. Shriver, and Rebecca Clausen. 2016. "Sustainability and Environmental Sociology: Putting the Economy in its Place and Moving Toward an Integrative Socio-Ecology." *Sustainability* 8 (5): 437.

Macekura, Stephen J. 2015. *Of Limits and Growth: The Rise of Global Sustainable Development in the Twentieth Century*. New York: Cambridge University Press.

McEvoy III, James. 1972. "Multi- and Interdisciplinary Research—Problems of Initiation, Control, Integration and Reward." *Policy Sciences* 3 (2): 201–08.

Misak, Cheryl. 2013. *The American Pragmatists*. New York: Oxford University Press.

Mitcham, Carl. 1995. "The Concept of Sustainable Development: Its Origins and Ambivalence." *Technology in Society* 17 (3): 311–26. https://doi.org/10.1016/0160-791X(95)00008-F.

National Academies of Sciences, Engineering, and Medicine. 2017. *Communicating Science Effectively: A Research Agenda*. Washington, DC: The National Academies Press. https://doi.org/10.17226/23674.

National Research Council. 2008. *Public Participation in Environmental Assessment and Decision Making*. Edited by Thomas Dietz and Paul C. Stern. Washington, DC: The National Academies Press. https://doi.org/10.17226/12434.

Neumayer, Eric. 2010. *Weak Versus Strong Sustainability: Exploring the Limits of Two Opposing Paradigms*. 3rd ed. Cheltenham, UK: Edward Elgar.

Ostrom, Elinor. 2009. "Beyond Markets and States: Polycentric Governance of Complex Economic Systems." Nobel Prize Lecture, Stockholm University, Sweden, December 8, 2009.

Ostrom, Elinor. 2010. "A Long Polycentric Journey." *Annual Review of Political Science* 13 (1): 1–23.

Ottaviani, Jim, and Leland Purvis. 2009. *Suspended in Language: Niels Bohr's Life, Discoveries, and the Century He Shaped*. 2nd ed. Ann Arbor, MI: G.T. Labs.

Roberts, J. Timmons, Julia K. Steinberger, Thomas Dietz, William F. Lamb, Richard York, Andrew K. Jorgenson, Jennifer E. Givens, Paul Baer, and Juliet B. Schor. 2020. "Four Agendas for Research and Policy on Emissions and Well-Being." *Global Sustainability* 3 (e3): 1–7. https://doi.org/10.1017/sus.2019.25.

Robinson, John. 2004. "Squaring the Circle? Some Thoughts on the Idea of Sustainable Development." *Ecological Economics* 48 (4): 369–84.

Rosa, Eugene A. 1998. "Metatheoretical Foundations for Post-Normal Risk." *Journal of Risk Research* 1 (1): 15–44.

Rosa, Eugene A., Ortwin Renn, and Aaron M. McCright. 2013. *The Risk Society Revisited: Social Theory and Governance*. Philadelphia: Temple University Press.

Steinberger, Julia K., J. Timmons Roberts, Glen P. Peters, and Giovanni Baiocchi. 2012. "Pathways of Human Development and Carbon Emissions Embodied in Trade." *Nature Climate Change* 2 (2): 81–85.

Stokes, Donald E. 1997. *Pasteur's Quadrant: Basic Science and Technological Innovation*. Washington, DC: Brookings Institution Press.

WCED (World Commission on Environment and Development). 1987. *Our Common Future*. Oxford: Oxford University Press.

Whyte, Kyle Powys. 2013. "On the Role of Traditional Ecological Knowledge as a Collaborative Concept: A Philosophical Study." *Ecological Processes* 2 (Article 7): 1–12. https://doi.org/10.1186/2192-1709-2-7.

Whyte, Kyle Powys, Joseph P. Brewer II, and Jay T. Johnson. 2016. "Weaving Indigenous Science, Protocols and Sustainability Science." *Sustainability Science* 11 (1): 25–32.

Worster, Donald. 2016. *Shrinking the Earth: The Rise and Decline of American Abundance*. New York, NY: Oxford University Press.

York, Richard. 2013. "Metatheoretical Foundations of Post-Normal Prediction." In *Structural Human Ecology: New Essays in Risk, Energy, and Sustainability*, edited by Thomas Dietz and Andrew K. Jorgenson, 19–29. Pullman, Washington: Washington State University Press.

III | Focusing Sustainability Education on Problem-Based Learning

8
Overcoming the Terrors of the Either/Or

ANN DALE

> The contemporary university is often characterized as working with colonized knowledge...The epistemologies of most peoples of the world, whether Indigenous, or excluded on the basis of race, gender or sexuality are missing. (Hall and Tandon 2017, 7)

THE CHALLENGES FACING MODERN SOCIETY are like never before; to name only two, climate change adaptation and mitigation and biodiversity loss are messy, wicked problems (Rittel and Webber 1973), what some refer to as super wicked problems. The natural world and human societies are now co-evolving (Norgaard 1994), and thus modern challenges are beyond any one sector, any one discipline, or any one government to solve (Dale 2001). They are scientifically complex and bio-complex, geographically unfocused, and embedded in landscapes that are rapidly changing, with multiple thresholds and tipping points that are essentially unknowable. Super wicked problems challenge our current post-secondary disciplinary structure and demand a new integrated sustainability science and collaboration between the humanities and the arts.

For the purposes of this chapter, and for my twenty-year research agenda, I define sustainability as the reconciliation of the ecological, social, and economic imperatives, and equitable access to these imperatives is essential for its realization (Dale 2001). My social location and background as a professor for twenty years, and now as director of the School of Environment and Sustainability at Royal Roads University, inform this chapter. I have taught both online and in-class courses for our MA and MSC students in the Master of Environment and Management program on inter alia sustainable community development, theories and stories of sustainable development, reconciliation and regenerative innovation, and leadership and sustainable development. I have supervised over one hundred students to successful completion of their theses, as well as several doctoral candidates. Previously, I was a senior associate at the University of British Columbia for a decade, helping to build the Sustainable Development Research Institute. All of these experiences have contributed to my ideas concerning the role and influence of post-secondary institutions in both educating and becoming living laboratories for the implementation of sustainability.

Most of the "big sticky questions" researchers and educators are facing are ultimately not necessarily those of scientific or managed origin; rather, they are about dealing with people and their diverse cultures, interests, visions, priorities, and needs (Norgaard 1994). Thus, climate change mitigation and adaptation, one of the big sticky questions, lies at the cross section of a lot of different fields, requiring expertise and model systems from diverse disciplines.

The structural challenges post-secondary institutions face are diverse: the disciplinary structure of the academy; the lack of integration within and between disciplines; the qualities that are valued in good research; the "suppository" relationship between the natural and social sciences; and the disincentives mitigating against inter- and transdisciplinary research teams. And innovative and critical scholarship is less likely to be funded or published in so-called "top drawer" journals (Brodie 2012, 14).

I would like to quote from a nineteenth-century French philosopher, Claire Démar, who greatly influenced my dissertation work and the coining of the phrase *solitudes, silos and stovepipes* (Dale 2001). For me, this quote describes the problematic structure of our knowledge systems, the terrorism of either/or when the world we live in is a world of "and." These unnatural separations and artificial dualisms keep us mired in the old problems and obscure the "real" solutions and social innovations needed to respond to the broad, horizontal, deeply interconnected challenges facing modern society. As Démar wrote in *Textes sur l'Affranchissement des Femmes (1832–1833)*,

> You proclaim two natures? Indeed tomorrow, depending on how many declare themselves to belong to the one or the other…You'll make one, perhaps involuntarily, predominate over the other; and soon we'll have a bad and a good nature, an original sin…you shall be the God and I shall be the Devil. (Démar 1976, quoted in Dale 2001, 17)

Unlike most other feminists of the nineteenth century who venerated difference and argued for the value of two natures on the basis of morality, Démar feared the authoritarian dynamics of what she called the classifications, the subtle and metaphysical distinctions by which humanity divides itself into a series of orders, classes, and types. And we have to question, as researchers, can we even ask the right questions, the "big sticky questions," given the dominant paradigm in which humans are perceived as apart from nature, rather than a part of nature? Are we responding to the immediate rather than the important (Toope 2010)?

Over three hundred years ago, Leibnitz suggested that university organization, in terms of faculties, impeded the expansion of knowledge across and beyond disciplines (Max-Neef 2005). Discipline terrorism arises when the disciplinary structure becomes so brittle and impermeable, rather than fluid and organic and permeable, that disciplines lose their functionality. And "of necessity, transdisciplinary work is based on disciplinary practice…it is complementary" (Klein 2004, 524). So, where do the solutions lie?

It is clear that the disciplinary structure of the academy no longer "fits" modern day challenges that, by definition, are cross-cutting, inter- and transdisciplinary, and intersectional. The lack of integration between the natural and social sciences is also a major barrier to the resolution of messy, wicked problems. And yet, intellectual homelessness (Golde and Gallagher 1999) is not valued, and the traditional disciplines have failed to embrace the sustainability imperative, forcing it to stay at the margins of their boundaries, with some permeability but no penetration.

Modern day problems demand new post-secondary systems that embrace inter- and trans-discipline research and teaching, enhance collaboration between the sciences and move to integrated science faculties, and stimulate intersections within disciplines.

Discipline and trans-discipline must be understood as complementary. The transit from one to the other, attaining glimpses from different levels of reality, generates reciprocal enrichment that may facilitate the understanding of complexity.

Transdisciplinarity, more than a new discipline or super-discipline is, actually, a different manner of seeing the world, more systemic and more holistic. (Max-Neef 2005, 15)

It is often at the "edges" that more innovation emerges, just as the richest ecosystems are those where the edges of two systems interact. Emergent bridging disciplines, such as systems biology and geo-medicine, must be given more space to accelerate the diffusion of knowledge more rapidly to wider audiences.

What are some of the fundamental literacies students need in this modern world? Trans-discipline skills need to be modelled and taught; they are not inherent, and demand an unprecedented degree of collaboration. They include interpersonal and intrapersonal skills (trust); cognitive skills such as differentiating, reconciling, and synthesizing (Lyall and Meagher 2012); communication across disciplinary, epistemic, and methodological boundaries; the scholarship of integration (Lattuca 2001); and independent critical thinking, self-organization, community service, and novel ways of sharing knowledge. And personal flexibility and openness; Wals and Schwarzin (2012, 13) argue for "a more systemic and reflexive way of thinking and acting, bearing in mind that our world is one of continuous change and ever-present uncertainty." This suggests that we cannot think about sustainability in terms of problems that are out there to be solved or in terms of "inconvenient truths" that need to be addressed. Instead, we need to think in terms of challenges to be taken on in the full realization that, as soon as we appear to have met the challenge, things will have changed and the horizon will have shifted once again.

Our Context

Higher education globally is growing, with 150 million students worldwide and projected to rise to over 260 million by 2025 (Maslen 2012). Two and a half million are internationally mobile, a number expected to grow to 7 million by 2020, and expenditure is already some $1.8 trillion and increasing at approximately 8 percent each year (Mulgan, Townsley, and Price 2016). Other major trends are the proliferation of social media and its ubiquitous use by younger generations, demanding dynamic new modes of teaching. Two other key trends are the rise of big data and growing sophistication of artificial intelligence, again demanding virtuous system innovation, which by necessity implicates post-secondary institutions.

At the conclusion of the United Nations Decade of Education for Sustainable Development (2005–14), the resulting *UNESCO Roadmap for Implementing the Global Action Programme on Education for Sustainable Development* still emphasized the importance of "designing teaching and learning in an interactive, learner-centred way that enables exploratory, action-oriented and transformative learning," as well as "rethinking learning environments—physical as well as virtual and online—to inspire learners to act for sustainability" (UNESCO 2014, 12). In spite of many exhortations about the need to change the capitalist system, question the global growth imperative, and move to a carbon neutral economy, the basic model of university education has not changed much since World War II and continues to reinforce and reward a feudal disciplinary structure of knowledge.

We are clearly facing multiple obstacles of an institutional, epistemological, and methodological nature (Darbellay 2015) that threaten the existing post-secondary structure. The nature of the changes needed is clearly transformative, and perhaps the best way to begin the change is at the edges moving into the centre, with a focus on curriculum design and delivery in and across disciplines. This demands a level of interdisciplinary understanding and collaboration between disciplines, as well as more dialogue, interaction, and negotiation than mastering, controlling, and expertise (Darbellay et al. 2014).

A Possible New Path: The Challenge-Driven University

One possible model for change in the short term is to adopt a challenge-based approach (Mulgan, Townsley, and Price 2016). This approach puts students up against messy, wicked problems and challenges for which there are no single right answers. Instead, students draw on many disciplines to solve them; they have to work in interdisciplinary teams; and they have to collaborate with organizations from the communities in which they are embedded. Since transdisciplinarity is fundamentally about "problem-solving" (Clark 2002) and "issue-driven research" (Robinson 2008), it brings political, social, and economic actors, as well as ordinary citizens, into the research process itself. Thus, the "problems" should be co-established with local governments, civil society leaders, and the private sector.

This model does build on the strengths of the disciplines and the benefits of working across disciplines rather than simply between them, for the indeterminate and boundary crossing nature of sustainability issues implicates all disciplines. By refocusing curriculum development on the challenges now facing modern day societies, post-secondary institutions will evolve to become more relevant to the co-creation of synthetic solutions that contribute to increased evidence-based decision making. More integrated decision making, grounded in the best science and evidence, is so necessary, given the fundamental changes that are necessary to achieve on-the-ground implementation and resolution of the many spatial justice (Soja 2003) issues now facing the world.

Royal Roads University (RRU), since its establishment in 1995, has been using some of the tenets of the challenge-based university articulated in the Wals and Schwarzin (2012) paper. In its cohort-based model in the Master of Environment and Management program, the emphasis is on problem-based case studies and an integrated Master of Arts and Science program. It also has the only distributed education model in Canada, that is, an iterative curriculum of three three-week, in-person residencies, each interspersed with online distance learning. The distributed model combines the best of both worlds—face-to-face knowledge sharing and leading edge online learning technologies.

Another RRU curriculum innovation is a new certificate program in sustainable community development, offered for the first time in

June 2017. Grounded in the principle of producing useful knowledge for decision makers, its design includes working with city officials to integrate local challenges as live case studies into its delivery. The blended certificate is composed of three sequenced courses. The first involves a three-month online course that shares the theory, practices, and tools that are now leading in sustainability epistemology. Building upon the four systemic properties of general systems thinking (Laszlo 1996), emphasis is placed on new and radical models of collaboration, integrated and creative decision making, climate change adaptation and mitigation, and diversity and regeneration (Dale 2001; Dale et al. 2014). In an in-depth examination of key issues, students gain a deeper understanding of the complexities involved in solving modern day "messy, wicked" problems (Paquet 1991) for which there is no single map or model, solution, or right answer.

A face-to-face residency then follows, in which the students select current sustainability challenges that have been identified in partnership, initially with the City of Victoria and subsequently with the City of Colwood. A combination of experiential learning and leading edge tools and techniques are applied to the resolution of social challenges identified by the partner cities. The students face their community of focus in a hands-on, highly engaging process.

In addition to the emphasis on real-world learning opportunities, a key feature of the residency is an emphasis on a multi-sensory experience of "being-in the-city" and learning on two levels, cognitive and affective. Sipos, Battisti, and Grimm (2008, 68) emphasize transformative education, where learning objectives are "organized by head, hands and heart—balancing cognitive, psychomotor and affective domains." Hart (2001, 7) encourages "an education of inner significance," where transformational experiences are more likely to occur when a link is made, and capacity is built between the interiority of the student and the external world. Consequently, place-based assignments, inter alia the "lay of the land," were created that allow students the space to reflect and "walk the city" during the residency.

At the end of the face-to-face residency, students present their cohort-based research plans to a community panel composed of city staff and RRU researchers. Following this critical feedback, they then work again online for another three months to complete their research and present their plans to elected officials and staff from the city. Interestingly, over half the students enrolled in the certificate course have gone on to enroll in a master's program at the university.

This model could be applied to university curriculum across the country, although it may face challenges if it is scaled to larger universities. It could be adapted by larger institutions, for example, by bringing together all undergraduate students from the natural and social sciences and the humanities, in the final year of their degree programs, into challenge research teams that cut across the disciplines. They could be convened in cohorts that deliberatively mix students from diverse disciplines around social challenges identified by political leaders in the communities in which they are embedded. These courses would emphasize both skills and knowledge practices, including research dissemination and communication, by presenting back to the community through workshops, to elected officials and staff, or at public community events. Ideally, the courses would be co-taught through interdisciplinary teaching collaboratives.

In summary, challenge-driven universities are deeply engaged in transdisciplinary research, and they teach integrative sciences, provide self-organized and blended learning choices, and, in the classroom, apply the co-creation of applied knowledge to challenge- and solutions-based learning. The institutions themselves should be living laboratories of sustainability, with operational and academic alignment, deeply embedded and contributing to the communities in which they live through their research.

Conclusion

Problem-solving curriculum grounded in real-world challenges that infuses sustainability knowledge within the disciplines will necessitate interdisciplinary teaching collaboratives across departments. This will require major structural change, both within universities and the granting councils, with associated values and incentives. What makes for a good researcher—the ability to persevere and adhere to individual research visions—sometimes mitigates against team collaboration. In addition to dealing with cross-university challenges in curriculum design, they will need to open up disciplinary silos, expand their boundaries to integrate sustainability knowledge, embrace the intersections and edges of research and teaching models, and transcend solitudes in the natural and social sciences, humanities, and the arts. How to sustain the best of disciplinary knowledge while at the same time being open to emerging new transdisciplinary knowledge and fields is a major question that post-secondary institutions face in this decade. As well, the nature of the "beast" requires novel university–civil society relationships and community engagement. More critically, the mechanisms and dynamics of current evaluation processes will have to be transformed to understand and reward inter- and transdisciplinary scholarship and research.

A caution, however. Transdisciplinary research is not necessarily a panacea if, as Gordon (2014) explains, it was the efforts to colonize reason that led to the generation of disciplines—transdisciplinarity, too, can be susceptible to decadence if it fails to bring reality into focus.

References

Brodie, Janine. 2012. "Social Literacy and Social Justice in Times of Crisis." Big Thinking Lecture Series. Canadian Federation of the Humanities and Social Sciences, 2012 Congress. Wilfrid Laurier University/University of Waterloo, May 30, 2012. http://www.fondationtrudeau.ca/sites/default/files/u5/social_literacy_and_social_justice_in_times_of_crisis_-_janine_brodie.pdf.

Clark, Tim W. 2002. *The Policy Process: A Practical Guide for Natural Resource Professionals*. New Haven and London: Yale University Press.

Dale, Ann. 2001. *At the Edge: Achieving Sustainable Development in the 21st Century*. Vancouver: UBC Press.

Dale, Ann, Robert Newell, Yuill Herbert, and Rebecca Foon. 2014. *Community Vitality: From Adaptation to Transformation*. Tatamagouche, Nova Scotia: Fernweh Press.

Darbellay, Frédéric. 2015. "Rethinking Inter- and-Transdisciplinarity: Undisciplined Knowledge and the Emergence of a New Thought Style." *Futures* 65 (January): 163–74.

Darbellay, Frédéric, Zoe Moody, Ayuko Sedooka, and Gabriela Steffen. 2014. "Interdisciplinary Research Boosted by Serendipity." *Creativity Research Journal* 26 (1): 1–10.

Golde, Chris M., and Hanna Alix Gallagher. 1999. "The Challenges of Conducting Interdisciplinary Research in Traditional Doctoral Programs." *Ecosystems* 2, no. 4 (July–August): 281–85. http://www.jstor.org/stable/3659018.

Gordon, Lewis. 2014. "Disciplinary Decadence and the Decolonisation of Knowledge." *Africa Development* 39 (1): 81–92.

Hall, Budd L., and Rajesh Tandon. 2017. "Decolonization of Knowledge, Epistemicide, Participatory Research and Higher Education." *Research for All* 1 (1): 6–19. https://doi.org/10.18546/RFA.01.1.02.

Hart, Tobin. 2001. *From Information to Transformation: Education for the Evolution of Consciousness*. New York: Peter Lang.

Klein, Julie T. 2004. "Prospects for Transdisciplinarity." *Futures* 36 (4): 515–26.

Laszlo, Ervin. 1996. *The Systems View of the World: A Holistic Vision for Our Time*. Cresskill, NJ: Hampton Press.

Lattuca, Lisa R. 2001. *Creating Interdisciplinarity: Interdisciplinary Research and Teaching among College and University Faculty*. Nashville: Vanderbilt University Press.

Lyall, Catherine, and Laura R. Meagher. 2012. "A Masterclass in Interdisciplinarity: Research into Practice in Training the Next Generation of Interdisciplinary Researchers." *Futures* 44 (6): 608–17.

Maslen, Geoff. 2012. "Worldwide Student Numbers Forecast to Double by 2025." *University World News*, February 19, 2012. https://www.universityworldnews.com/post.php?story=20120216105739999.

Max-Neef, Manfred A. 2005. "Foundations of Transdisciplinarity." *Ecological Economics* 53, no. 1 (April): 5–16. https://doi.org/10.1016/j.ecolecon.2005.01.014.

Mulgan, Geoff, Oscar Townsley, and Adam Price. 2016. "The Challenge-Driven University: How Real-Life Problems Can Fuel Learning." NESTA. March 2016. https://media.nesta.org.uk/documents/the_challenge-driven_university.pdf.

Norgaard, Richard B. 1994. *Development Betrayed: The End of Progress and a Co-evolutionary Revisioning of the Future*. London: Routledge.

Paquet, Gilles. 1991. "Policy as Process: Tackling Wicked Problems." In *Essays on Canadian Public Policy*, edited by Thomas J. Courchene and Arthur E. Stewart, 171–86. Kingston: Queen's University School of Policy Studies.

Rittel, Horst W.J., and Melvin M. Webber. 1973. "Dilemmas in a General Theory of Planning." *Policy Sciences* 4 (2): 155–69.

Robinson, John. 2008. "Being Undisciplined: Transgression and Intersections in Academia and Beyond." *Futures* 40 (1): 70–86.

Sipos, Yona, Bryce Battisti, and Kurt Grimm. 2008. "Achieving Transformative Sustainability Learning: Engaging Head, Hands and Heart." *International Journal of Sustainability in Higher Education* 9 (1): 68–86.

Soja, Edward. 2003. "Writing the City Spatially." *City* 7 (3): 269–80.

Toope, Stephen J. 2010. "A (Not 'The') UBC Response to Nigel Thrift's Questions on Global Challenges and the Organizational-Ethical Dilemmas of Universities." *GlobalHigherEd*, December 5, 2010. https://globalhighered.wordpress.com/2010/12/05/a-ubc-response-to-nigel-thrift/.

UNESCO (United Nations Educational, Scientific and Cultural Organization). 2014. *UNESCO Roadmap for Implementing the Global Action Programme on Education for Sustainable Development*. Paris: UNESCO. https://sustainabledevelopment.un.org/content/documents/1674unescoroadmap.pdf.

Wals, Arjen E.J., and Lisa Schwarzin. 2012. "Fostering Organizational Sustainability through Dialogic Interaction." *The Learning Organization* 19 (1): 11–27.

9
Sustainability Education
A Dance Between Knowledge and Experience

SHIRLEY M. MALCOM

Growing up Black and female in the 1950s and 1960s segregated South, the concept of sustainability was not on our radar. The economy of the region was based on accessing and using resources with little thought of consequences. But looking back, I now see that, at least for the African American community, sustainability was something we lived and understood. We wasted little, recycled everything, and understood the trade-offs in our lives—our livelihood and economy on the one hand, and our environment and health on the other. Developed as a city in post-Civil War Alabama, Birmingham was notable as a site where there were significant deposits of iron ore, coal, and limestone—the main ingredients for making steel. And into the steel mills and blast furnaces came the migrants—cheap, non-unionized Black labour from the surrounding rural areas.

My hometown borrowed its name from the major industrial centre, Birmingham, England, and the city became known as the "Pittsburgh of the South," sharing its economic reliance on iron and steel and reaping the environmental degradation that came with it. This city also became the epicentre of one of the greatest social movements of our time—the battle for civil rights for Blacks in the United States.

Decades later, the steel mills and mines closed. But as with other cities built on this economic base, such as Pittsburgh, the environmental damage remained. Now came the need to reclaim the environment—the land, the water, and the air. I now realize that it was these experiences, as much as anything, that left me open to a need to understand the systems upon which life on Earth depends. I have also come to understand more about the relationship between civil and human rights and sustainability.

I OFFER THIS OPENING REFLECTION to connect my positionality to how I came to understand sustainability. My formal education is in the life sciences, framed by ecology, and my doctorate is in ecology. The living systems perspective that was born in my childhood has informed my life and my life's work. As an educator, I have been aware of the power of ecology to enable us to connect with communities as well as with students. As a social justice advocate, I have come to understand the impact on marginalized communities of decision making related to the economy and the environment, often posed as trade-offs.

Sustainability as a concept crosses many disciplines in the physical, computational, biological, earth, social, and behavioural sciences, and it incorporates and builds on ideas where many existing areas of scholarship reside. Yet the concept of sustainability transcends all these areas. I view this as articulated in the Brundtland report: "Sustainable development is development that meets the needs of the present without compromising the ability of future generations to meet their own needs" (WCED 1987, 43). In this chapter, I raise issues of the environment, the economy, and equity.

If we are working to prepare researchers and practitioners for future work in sustainability, what should we do? How would we go about developing an educational program focused on sustainability? For example, what would be the balance between "book learning" and practical experience? If we are to consider how universities relate to sustainability, we must ask whether the way they are organized can support such education, and what might need to be modified to accommodate this topic as an area of scholarship and/or action. What does the proposed "body of work" look like for sustainability education? What experiences should be provided to students? Beyond technical knowledge, what skills and mindset do students need? If I were an advisor for a program, what would I suggest that students' portfolio of study and experiences include?

We could name many different topics, courses, and programs that we might want students to have, across many different fields, within and beyond my areas of interest in science, technology, engineering, and mathematics (STEM). But the bottom line is that the inherent multidisciplinary nature of sustainability makes it hard to nail down, hard for an advisor to advise on, and hard for a student to consider: what is the core or foundational work a person might need to contribute to sustainability research and/or practice, and what is the *more*, or elective work, that would relate to a particular student's interests and strengths or local contexts? And beyond curriculum, what skills would students need to acquire, and what experiences might be provided to support acquiring these skills?

It was pointed out to me that we now often ask college students, "What's your major?" But for sustainability education, it may be more apt to ask, "What's your problem? What do you want to understand? Which aspect of this very big concept are you taking on? What is the lens through which you want to view or contribute to sustainability?"

One lens I use to view sustainability is that of ecology, the area of focus for my doctoral research and my foundational focus of study since my undergraduate years at the University of Washington. But the biology of my undergraduate and graduate days has moved on: the Human Genome Project, epigenetics, and the Human Microbiome Project all require an expanded vision of ecology—for example, to include each human's own collection of microorganisms. Such considerations require an expanded frame to look at the systems around us.

Another lens I use to view sustainability is that of human rights and social justice. In exploring options in problem solving in the context of sustainability, how do we ensure that the advantages of proposed solutions are widely shared and that the disadvantages are not differentially imposed, especially on marginalized groups? How does one learn not only to "do things right" but also to "do the right thing?"

The Challenge of Institutional Arrangements

Colleges and universities have used many different strategies for organizing themselves when addressing cross-cutting areas of scholarship and practice, such as with sustainability. A report I co-authored thirty years ago outlined how institutions navigated programs such as those designed to increase the presence of women and underrepresented minorities in science and engineering, another cross-cutting theme (Matyas and Malcom 1991). Although the framework outlined in that report focused on a different cross-cutting theme, that of diversity in STEM, it mirrors somewhat the approaches often seen when tackling interdisciplinary topics: from isolated offerings scattered across the institution, largely driven by the personal interest and commitment of a faculty member, to the aggregation of courses perhaps within an organizational unit, to loose coordination across programs and units and then the creation of centres, to, finally, structural reform.

FIGURE 9.1: Evolution of Sustainability Programs

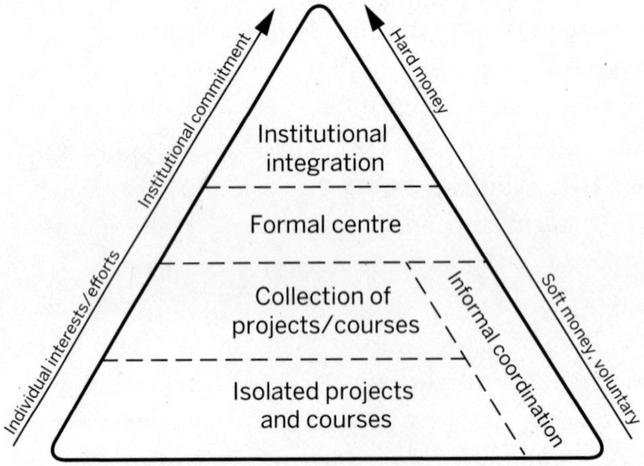

[Triangle diagram with tiers from bottom to top: Isolated projects and courses; Collection of projects/courses; Formal centre; Institutional integration. Left side labeled "Individual interests/efforts" to "Institutional commitment". Right side labeled "Soft money, voluntary" to "Hard money". Dashed arrow labeled "Informal coordination".]

Credit: Created by Shirley M. Malcom (adapted "Model for the Evolution of Intervention Programs" figure in Matyas and Malcom 1991, 144).

Isolated Courses

Isolated courses focused on sustainability often appear on campuses, driven by the research interests or activities of individual faculty members. These are valuable but may disappear if the particular champion for that topic leaves the institution. It may also be hard to find among myriad course offerings. Online courses offer additional options to students. Coursera (2020), for example, offers some 129 courses in different languages under the title of environmental science and sustainability.

Collections of Coursework
For very practical reasons related to the organization of institutions, it may be difficult to fully consider the interdisciplinary space of sustainability. When we begin to look at any course of study that lies beyond a single field, there is a great temptation to add more courses to "fill in" traditional disciplines. The result of that approach has often been the "overstuffed curriculum." While packing in more discrete bits of information, it does not necessarily provide access to an integrated whole. That is, students are given the words, but not the plot, the context, or the story. So, the central message of sustainability may be lost. In addition, with a focus on filling in with coursework, the practical/practice concerns may be totally omitted.

Aggregation Across the Theme
There may be formal or informal coordination of discrete projects that are distributed across the institution. The Cornell Atkinson Center for Sustainability (2019) listed 886 courses across the university that were sustainability focused or related, with individual faculty and/or departments opting onto the list. The challenge of cobbling together a coherent picture remains for each student. A personalized program may be possible for savvy students, but it is not clear how, for example, a first-generation college student would know how to navigate across these offerings without assistance in defining a destination and a pathway that also satisfies general institutional requirements.

Special Programs or Centres
Some institutions take the step of creating a program that has specific goals and that may carve out some aspect of sustainability. For many institutions in the US, the availability of funding for a centre or institute often paves the way for multidisciplinary programs to get their start.

There are both advantages and disadvantages to employing this strategy. On the plus side, donor interest can be a driver to obtain the support of institutional leadership; existing faculty from adjacent interest areas can often be recruited into a centre; and soft money support provides an opportunity to build a constituency and to nurture student interest. On the minus side, the structures themselves do not change; faculty members' primary home often remains within the department where their "lines" are located, their work is evaluated, and the basis for evaluating their scholarship continues to reside. And "soft money centres" sometimes disappear when funding disappears, or they remain as an organizational shell without resources to support action.

Systemic Approaches
The hardest task is to build sustainability into the fabric or mission of the institution in such a way that it is supported by institutional funds and has dedicated staff—where the scholarship and service of faculty are recognized and counted within the system of rewards, a coherent vision of the work is articulated, and the experiences and learning opportunities are designed to support that vision. Systemic, or structural, approaches may also be important for political reasons since this may force a reconciliation of the issue of the faculty "home" and faculty "work." As long as a faculty member's home remains within a department and that person's work is evaluated there, then the standards of scholarship reside within the department.

Student Outcomes

What should the students look like who emerge from programs in sustainability? There need to be goals and expectations articulated for those who complete these programs. We should be able to say what we want students to know and be able to do, and what mindset and skills we want them to have. Only by thinking about student outcomes can we begin to discuss the knowledge and experiences that need to be provided to students by institutions, so that they are enabled to contribute to the scholarship and practice of sustainability. Core concepts and competencies can be taught using many different frames from many disciplinary areas.

The challenge in preparing students for multidisciplinary fields is addressing the need for students to have a deep knowledge of at least one area (a strong base), even as they obtain a working knowledge of other related fields. In addition, there are skill sets that are vital to sustainability: written and oral communications, global awareness, cultural competence, working in teams, and quantitative and computing skills.

Programs that focus on sustainability need to aid in the development of so-called "T-shaped" people. This term has been used extensively in the business community to describe their approach to talent management, especially in the kind of collaborative culture that sustainability also requires. In an interview by *Chief Executive* magazine with Tim Brown, the CEO of the design consultancy firm IDEO, Brown talked about T-shaped people:

> T-shaped people have two kinds of characteristics, hence the use of the letter "T" to describe them. The vertical stroke of the "T" is a depth of skill that allows them to contribute to the creative process. That can be from any number of different fields: an industrial designer, an architect, a social scientist, a business specialist or a mechanical engineer. The horizontal stroke of the "T" is the disposition for collaboration across disciplines. It is composed of two things. First, empathy. It's important because it allows people to imagine the problem from another perspective—to stand in somebody else's shoes. Second, they tend to get very enthusiastic about other people's disciplines, to the point that they may actually start to practice them. T-shaped people have both depth and breadth in their skills. (Hansen 2010)

Brown spoke of the characteristics of empathy, such as putting oneself in the shoes of others, using listening skills, communicating, and building on the ideas of others. He noted that, at the end of the day, T-shaped people have to "get things done." In educating people for settings where the focus is on solving a problem, we can learn from other fields.

A major challenge with sustainability is the size and expansiveness of the knowledge base that can be required to solve problems. Thus, there are advantages to having added areas of depth, such as can be seen with "π (pi) shaped" skills and "comb-shaped" skills (Dawson 2013). These added areas of expertise will likely grow with time, but the base for that later growth is in the core work and in the emphasis on collaboration and lifelong learning.

Faculty Perspectives

What happens to a faculty member who decides that they have an interest in working in an interdisciplinary area? It depends. Is this an established (tenured) researcher who wants to shift fields, or a new hire emerging from an interdisciplinary graduate program? Is there an established community with organizations, conferences, and journals? An emerging community? What is the position of that person relative to the institutional arrangements? What expectations are being articulated by their departmental home?

When considering an area such as sustainability, it is clear that working in this space requires a blurring of the lines across teaching, research, and service. At its best, the problems come from the community or client, and engaging students in that work is part of the active learning that helps them acquire the skill sets that go beyond technical competence—that "cross the 'T.'" New knowledge will emerge in the process of solving the problems. It is likely that such faculty will need to become students as well, as they interact with peers in other fields, with students who may bring diverse perspectives, or with local people or practitioners. Again, it involves the interaction of knowledge and experience—one's own and that of others.

Finding and engaging with students from diverse populations will be important to doing one's work and capitalizing on the perspectives that derive from a diverse and inclusive team. This will mean resisting the tendency to duplicate a faculty member's background and experience or to find people with whom one is "comfortable." A faculty member might be required to recruit students to a program, rather than just sort from among those who apply. And once these diverse students are attracted to one's program, it will be important to let them help shape the questions asked, to avoid placing one kind of knowledge as superior to another, and to recognize and honour the local knowledge systems they might bring with them.

How do faculty members find collaborators beyond their own disciplinary area and possible clients for their work? Often this is best accomplished by publicly sharing one's knowledge, interests, and "problems" in accessible presentations aimed at non-specialist audiences.

One example of this is the Canon National Parks Science Scholars Program, which provided dissertation research support to enable graduate students to complete work leading to their PHD in research areas critical to national parks. The result of a partnership from 1997 through 2007 among Canon U.S.A, Inc., the National Park Service, and the American Association for the Advancement of Science (AAAS), these scholarships cut across engineering/technology and the physical, biological, social, and cultural sciences, and they required—in addition to the completion of a thesis—a public presentation of the research to demonstrate its relevance to the management of the parks. These presentations conveyed the "so what" of the research that had been undertaken. While the research was valuable in its own right as it generated fundamental knowledge, the applications of the work provided added value. But these kinds of programs are rare. The program drew on generous support from a private sector funder whose values and interests in sustainability were aligned with those of the partners. The attitudes of corporation leaders, as well as corporate business practices, provided consistent messages around sustainability—they "walked the talk."

Conclusion

At the end of the day, universities are likely to struggle to find a model to seamlessly incorporate sustainability into the fabric of their structures. While they are billed as places where innovation can flourish, universities are also inherently conservative places. Their structures and traditions can be barriers to any field that does not conform to the existing architecture. This includes thinking about what expertise looks like, how we reward or punish "border crossing," and where the home for interdisciplinary work resides, as well as identifying the sources of support. Sustainability does not just live in the STEM space. So, how can one support work that also includes areas such as gender and race studies, civil rights and social justice, history, humanities, and the arts? The challenge of supporting interdisciplinary scholarship and scholars requires next generation thinking that values and takes advantage of knowledge and experience.

What will need to change for higher education to also "walk the talk?" Interestingly, I advise the same kind of "full body scan" to support sustainability research and education that I would advise to realize diversity, inclusion, and social justice goals. We will need a rethinking of scholarly work and the engagement of research—in practice, thinking, and mindset—as an element of good teaching, and a consideration of service as necessary to both. To date, many of the efforts to reinvent education or research have occurred with the establishment of new institutions. Examples of these include the Olin College of Engineering, the Picker Engineering Program at Smith College, and the Janelia Research Campus of Howard Hughes Medical Institute. In all these cases, many parts of the systems that "bake in" the traditional education and research structures were removed. But it is rare that resources at the levels required are available to start anew. We must figure out how to reinvent, renew, reconfigure, and reimagine existing institutions, to remove the constraints that restrict our movement toward the transformation in higher education that can support excellence in sustainability scholarship and education.

Today's Birmingham? The steel mills are gone; the blast furnaces no longer light up the night sky. There are some manufacturing jobs from automotive companies that have moved to the southern US to take advantage of non-unionized labour and tax incentives. The university is a major employer and source of business innovation, but only for those who are well educated. Birmingham has a smaller population today than when I left in 1963, largely due to "white flight" from the city to the surrounding suburbs. Unemployment is much higher than the national average.

In contrast, Pittsburgh has chosen a different path, with sustainability at its core. With support from a vibrant philanthropic community, it is reinventing and reimagining itself as "a region forged in steel" turning to a sustainable future. Its P4 efforts (people, planet, place, performance) are anchored in the need to focus on economic and social equity. And its universities are clearly a major part of the equation. As it cleans up its waterways, sustainably develops its restored brownfields, and builds up its "green" economy, Pittsburgh will need its universities and colleges to be active partners. And higher education can only achieve this by looking outward as well as inward, to the contexts in which the colleges and universities are based. Their futures are linked, yielding incredible opportunities to show that transformation is not only desirable but also possible.

References

Cornell Atkinson Center for Sustainability. 2019. "Cornell Sustainability Courses." Accessed May 28, 2020. https://www.atkinson.cornell.edu/education/curricula/.

Coursera. 2020. "Environmental Science and Sustainability." Accessed July 10, 2020. https://www.coursera.org/browse/physical-science-and-engineering/environmental-science-and-sustainability.

Dawson, Ross. 2013. "Building Success in the Future of Work: T-Shaped, Pi-Shaped, and Comb-Shaped Skills." March 21, 2013. https://rossdawson.com/building-future-success-t-shaped-pi-shaped-and-comb-shaped-skills/.

Hansen, Morten T. 2010. "IDEO CEO Tim Brown: T-Shaped Stars: The Backbone of IDEO's Collaborative Culture." *Chief Executive*, January 21, 2010. https://chiefexecutive.net/ideo-ceo-tim-brown-t-shaped-stars-the-backbone-of-ideoaes-collaborative-culture_trashed/.

Matyas, Marsha Lakes, and Shirley M. Malcom, eds. 1991. *Investing in Human Potential: Science and Engineering at the Crossroads*. Washington, DC: American Association for the Advancement of Science (AAAS). https://www.aaas.org/sites/default/files/s3fs-public/reports/Investing%2520in%2520Human%2520Potential_Science%2520and%2520Engineering%2520at%2520the%2520Crossroads%25201.pdf.

(WCED) World Commission on Environment and Development. 1987. *Our Common Future*. Oxford, UK: Oxford University Press.

IV | Cultivating Civic-Mindedness,
   Deliberative Dialogue, and
   Pathways toward the Public Good

10

Cultivating Courage in an Increasingly Complex, Divided World

TODDI A. STEELMAN

I WOULD LIKE TO POSE A VISION: an open, diverse society where university communities strive to educate students to wrestle honestly and fully with the complexities associated with addressing the globe's most challenging problems, such as race, terrorism, water security, food security, energy sustainability, poverty, immigration, education, and all the many other challenges worthy of our attention. As educators, students, and leaders, we have to understand more completely how we as practitioners of sustainability science and education fit into the collective mission of universities in our society. Sustainability, from my vantage point, is similar to "flourishing"—where the focus is on nurturing possibility for all humans and other life on earth (Ehrenfeld and Hoffman 2013).

I approach sustainability from the perspective of an educator with a thirty-year career that has moved from teaching as a professor in the classroom to administering as a dean of both small and larger schools of the environment—one in Canada and one in the United States.[1] As a teacher, I have instructed dozens of classes at four different academic institutions to thousands of students, including mentoring undergraduate, masters, and doctoral students through their intellectual journeys.

My research has focused on the human dimensions of environmental and natural resource policy, often at the community level. I investigate these issues through the lens of institutions, decisions, and behaviours. I firmly believe in the role of higher education as a transformative force in the world and want us to do a better job of fulfilling this important role.

Sustainability problems are complex because they are defined by uncertainty and conflicting value demands (Funtowicz and Ravetz 1995; Rittel and Webber 1973). Therefore, university communities need to become more comfortable with the uncertainties and conflicting values associated with these problems. Our role, our mission, and our purpose should be to help identify common interests and help secure the common good where we can. At a minimum, faculty and administrators have a responsibility to help identify and articulate where our cultural, economic, and political fault lines lie, so that society can more constructively deal with them. At this time in our history, we desperately need more light, not more heat. The words of Dr. Martin Luther King, Jr. (1963) resonate today: "We are confronted with the fierce urgency of now."

What Stands in Our Way?

Three key trends are important to consider as university leaders, faculty, and students contemplate the future of sustainability science and education. I describe each of these to suggest ways that we can remove the barriers these trends have imposed to more constructively address sustainability challenges. These are: declining trust in social institutions, a growing divide between those who are benefiting from a globalized economy and those who are not, and increased skepticism of expertise. Universities have essential roles to play in addressing these troubling trends. I would argue that sustainability science has an especially important role to play.

Distrust in Social Institutions

Higher education is under assault, but it is not an exception. Many institutions have faced a collapse in the trust society places in them—government, media, big business, the criminal justice system, and banks, among others. In many corners of the world today, there is a prevailing sense that institutions are failing the people they are meant to serve. A key priority for universities is to preserve institutions of higher education as places that are worthy of public trust. Practitioners of sustainability science are well positioned to earn that trust as agents who can demonstrate the continued relevance of higher education to the critical everyday problems people face.

Growing Economic Divides

New technologies and global trade agreements have transformed the economic opportunity structure across demographics. A slow-moving economic earthquake has created a fissure in the tectonic plates of socioeconomic expectation and reality. We are dividing into a world of those who believe in globalism and those who are strongly opposed to it—mostly because those who oppose it do not see themselves benefiting from it. Universities need to do their part to bring along all of society, not just the privileged few who can afford the traditional model of tertiary education. Sustainability science rests on three pillars—economic, ecological, and social. In short, sustainability science practitioners need to enhance our social equity mission, especially as it relates to those who are being left behind.

Credibility Gaps

"Expert" voices of authority no longer have the credibility they once did. With good reason. Witness the failures of experts, pundits, and the media in the botched predictions related to the United Kingdom's vote to leave the European Union or the 2016 US presidential election. Science no longer occupies an unchallenged role in informing how we make decisions or policy (Ascher, Steelman, and Healy 2010). Other sources of information are increasingly seen as legitimate competitors to "experts," such as information available through asking a Twitter, Instagram, or Facebook community, surveying a group of friends, or searching the internet (Tett 2016). Additionally, those with more authority are no longer automatically seen as more trusted. There are growing trends of people trusting "people like me" who move in social circles that source like-minded information. This shift from vertical axes of trust to more horizontal axes of trust is the norm among up-and-coming generations (Tett 2016). Increasingly, the only way to have credibility when competing with so many diverse sources and modalities is to engage directly with the communities where we want our expertise recognized. This means engaging in place-based, community-engaged research, where salience, legitimacy, and credibility can be established through knowledge co-production and collaboration. In other words, this is a model of transdisciplinary sustainability science.

How Do We Move Forward?

Universities need to be more courageous. Our faculties shrink away from the controversial and stay on the politically safe path. We need to find ways to increase trust in the public good, toward which higher education can contribute. We need to make higher education learning opportunities and degrees accessible and relevant to a larger swath of the population. We need to demonstrate our credibility with greater collaboration and engagement with the people who stand to benefit from sustainability educators' work. As leaders in our institutions, we need to use our authority to create the spaces for the extraordinary.

Space for What?

Increasingly, it seems like universities are creating spaces where we narrow the conversation. "Safe space" has become code for a kind of political correctness and orthodoxy that is corrosive to one of the most treasured values of academia—freedom of expression. We need to ask ourselves, freedom of expression for whom? For what purpose? Universities have long been places where freedom of speech and academic expression have been valued and practiced as part of an essential part of a functioning democracy.

Ultimately, politics is about the shaping and sharing of values and how to put into practice those values that we cherish (Lasswell and McDougal 1992). If universities do not create spaces where we can discuss our values—especially the most controversial ones—then we lose capacity to function as a democracy. We witnessed the very erosion of these values in Hungary, where the 2016 government passed legislation to close down Central European University, in part due to CEU's "open society" mission. This should be a wakeup call to all of our institutions of higher education across the West. Too often we are creating spaces and opportunities for people to retreat into their cultural or ideological foxholes.

As university leaders and faculty, we need to ask ourselves: what creates resilience? Resilience in individuals is the capacity to bounce back positively from challenging experiences (Masten et al. 2009). Universities should be training grounds for students to encounter different viewpoints, ideas, and people and learn how to deal with them constructively. Critically and respectfully engaging in why we disagree are the cornerstones of vibrant discourse in a democracy. Our classrooms have become places where we shy away from disagreement and seek consensus all too readily. What would happen if universities valued the diversity of ideas as much as we value the diversity of people? We need to create opportunities in classrooms, symposia, and lectures, and with invited speakers and panels to engage in the plurality of thought. In the sustainability sciences, this means we need to engage with the nuclear industry and its proponents if we want to have a conversation about sustainable energy. We need to invite those who believe in the power of genetically modified organisms if we are to fully consider the spectrum of food security. We need to question why we typically do not want to respect, hear, or understand these perspectives.

Why? Because we all have an obligation to listen. We do not have to agree or even come to a consensus. One of the strengths of a vibrant democracy is the ability to listen to and hear each other. Understanding that a perspective is different from yours and why it might be important to someone else is a big step in creating empathy and a search for common good, common sense solutions. If we have stopped listening, then we are really lost. And increasingly, it seems universities are more tolerant of the spaces where we do not have to listen to anything else other than what we already believe. So, as leaders, faculty, and students, we need to do a better job of creating spaces where we can listen to, hear, and engage in vibrant, diverse discussions about the complex, messy topics that confront us as a society. We need to develop the critical analytical skills to hear these differences, cultivate respect, and disagree civilly.

We need to do this in our classrooms, faculties, departments, schools, colleges, and our universities as a whole. Universities have multiple platforms where we can create these spaces. This means we need university faculty who are trained and experienced in dealing with conflict, facilitation, and mediation—qualities not typically found in job descriptions.

Space for Whom?
Consider who will be coming through our doors in the next decade and the implications for how universities prepare them to deal with sustainability challenges. We know that members of Generation Z (considered by some to be those born 1995-2010) are very comfortable with mobile technology and social media, and have lived through uncertain times punctuated by terrorist activity, shootings, and global recession (Segran 2016). They have always had cell phones. Their parents have experienced growing income gaps, debt, and a shrinking middle class (Williams 2015). The ability to afford college is a predominant concern, and the value of higher education is questioned by these students because they are paying more for it (Segran 2016).

To serve these students well, universities will need to be ready to adapt to their educational needs. To accommodate them as leaders and faculty, we desperately need to get out of our own heads and not expect them to be just like us. We say, "When I was in school…," "When I did my degree…," and Kids today…"—but look around. They do not need to adapt. We need find ways to be relevant to this group of students, so we can effectively convey our lessons to them. This includes understanding what kinds of learning modalities will best fit their and our collective needs.

Information literacy—or the ability to efficiently and effectively access information, evaluate it critically, and use it accurately—is crucial for this generation of students (Geck 2007). From the era of the Trump administration, we have learned of the dangers of "fake news" and "alternative facts" and the corresponding need to create learning opportunities for our students. They need to be able to sort through all the information at their fingertips and develop critical analytical skills for discerning fact from fiction in all they encounter, and we must provide opportunities for them to synthesize what they are learning from multiple diverse sources and to learn how to communicate effectively to multiple audiences. We will need to lead by example. This also means educating ourselves.

Teaching this next generation of students will call into question the very definition of what it means to be a professor. While "experts who will profess" will remain important, universities will also need skilled maestras and maestros who can compose, choreograph, orchestrate, and direct conversations that allow for a full range of ideas, technologies, and knowledges to be brought to bear in a pedagogical symphony.

What Can We Do Now?

There are three areas we can focus on to create more space for universities to be courageous in the face of an increasingly complex and seemingly divided world. These include creating supportive institutional incentives, developing creative and holistic curricula, and hosting creative spaces for controversial conversations.

Create Institutional Incentives for Community-Engaged Scholarship
If we want faculty and students to engage with communities to increase credibility and legitimacy, then we also need to create incentives to reward faculty who engage in this kind of scholarship. The predominant model of scholarship rests on independent discovery as the paragon of success. We need to work within our universities and colleges to create incentive structures for faculty to reach beyond the ivory tower into the areas where faculty can learn from communities and communities can learn from faculty. This is a transdisciplinary model that embraces knowledge co-production to define a problem and its potential solutions (Lang et al. 2012; Miller et al. 2014; Steelman et al. 2015). To effectively move in this direction, a systems-based approach is needed. This approach would align individual actions, departmental and college structural incentives and supports, as well as the broader external scholarship culture to encourage a shift toward recognizing and rewarding community-engaged scholarship. Universities are incredibly risk averse institutions. So, leaders need to think simultaneously about how to change institutions, which by their nature are enduring, to be something different. This means aligning incentives to mitigate risks for faculties, departments, colleges, and universities so they can engage in change.

Develop Holistic Curricula
Sustainability programs need curricula that imagine desired learning outcomes and design from those outcomes to establish coursework and

experiences that achieve those outcomes. Too often we are content with a piecemeal approach where curricula are patchworked together from courses faculty want to teach, and we disregard the overall learning objectives and how we can best achieve them (Clark et al. 2011a; Clark et al. 2011b). Leaders and faculty do not want to rock the boat. We need to rock the boat. We need to emphasize engaged learning that is experience based and gives back to communities. Working on real-world problems allows students to address the concrete, maddeningly complex details that books cannot provide. It moves us into a less ideological space. This means we need faculty and students to come together with a vision of what this would mean.

We need to do a better job of sharing examples and stories about how this is achieved. For instance, when I was executive director of the School of Environment and Sustainability at the University of Saskatchewan, our students began with a one-week immersive field skill experience working with farmers in the Redberry Lake Biosphere Reserve (Kricsfalusy, George, and Reed 2018). The students created a social and biophysical data set based on helping farmers become more sustainable in their practices, thus enabling the students to understand the complexities associated with farming. This provided the basis for the rest of the semester, in which the students analyzed the data and created a report that they then presented back to the farmers at the end of the term. Cumulatively, a longitudinal data set is being created for the farmers and the Biosphere Reserve to understand longer term trends and inform decision making. This field skills and data analysis course feeds into the two remaining core courses, which focus on theories and practices in sustainability and tools and applications in complex, sustainability problem solving. Additional electives are available, depending on the students' chosen path. Seven graduate attributes (learning outcomes) structure the entirety of the elective and core curriculum, including: (1) Think holistically with ethical intent; (2) Deeply understand sustainability; (3) Integrate a range of perspectives and ways of knowing; (4) Develop ambassadors for sustainability and agents of change; (5) Advance research expertise; (6) Demonstrate collaborative, leadership, and professional skills in knowledge sharing; and (7) Cultivate a substantive area of expertise in keeping with the student's program of study.

Host Controversial Conversations
People within and outside our campuses are hungry for conversation on important topics. Universities can provide constructive opportunities for diverse communities to come together and facilitate conversation. At my former institution at the University of Saskatchewan, a group of individuals external to the university showed up at the university senate to have a conversation about uranium mining and its impacts on First Nations. They were turned away because the senate was not the appropriate venue for this exchange, but these restrictive (and seemingly anti-democratic) rules are difficult to comprehend to those external to the sometimes byzantine ways of a university. In response, the chancellor created a completely separate forum for this discussion to take place. Four speakers, who could address different facets of this complex topic, were identified to speak for 10 minutes each. Topics ranged from the relative benefits and costs of the uranium mining industry for Indigenous Peoples in Saskatchewan, to the advantages and disadvantages of nuclear power as an energy source relative to other renewable sources, to the pros and cons of nuclear medicine. The audience was invited to submit cards with questions to a moderator during the talks, and the moderator then posed them to speakers to ensure they all received questions and that the questions were respectful. One outcome of this forum was that those who felt they had previously been silenced had the opportunity to be heard. Different viewpoints were expressed, and I believe most people walked away having learned something important about a perspective other than the one with which they arrived. The balance of the speakers, structure of the forum, and facilitation of the question-and-answer period all contributed to the success of this effort.

The lesson for other institutions is to have courage and step into having the conversation so we can build empathy or at least erode the discord. Structured discussion series that embrace the full complexity of trading off environmental, social, and economic values associated with topics such as climate change, feeding the planet, realizing reconciliation with Indigenous Peoples, and creating realistic energy futures could make universities more relevant, especially if such topics are made concrete in place-based contexts. Skillful facilitation can contribute to finding the common ground where we are less divided, while recognizing important differences and the values they represent.

Conclusion

It is time for us to be brave, have courage, and step up to create the places and spaces for conversation, learning and listening so we can more effectively address the very real and pressing sustainability challenges we collectively face. The uncertainty and values conflict that typify complexity will not go away. Our only option is to learn to live with uncertainty and become more skillful in addressing our conflicting values. The problem-oriented, transdisciplinary, solutions-based approach imagined in sustainability science is one path forward to potentially soften the hard edges of ideology, demonstrate the relevance and trustworthiness of higher education, and validate the importance of expertise. Practitioners of sustainability science are perfectly poised to play an essential role in cultivating this courage in our increasingly complex world.

Note

1. At the time this chapter was written, I was executive director of the School of Environment and Sustainability at the University of Saskatchewan.

References

Ascher, William, Toddi Steelman, and Robert Healy. 2010. *Knowledge and Environmental Policy: Re-Imagining the Boundaries of Science and Politics*. Cambridge, MA: MIT Press.

Clark, Susan G., Murray B. Rutherford, Matthew R. Auer, David N. Cherney, Richard L. Wallace, David J. Mattson, Douglas A. Clark, Lee Foote, Naomi Krogman, Peter Wilshusen, and Toddi Steelman. 2011a. "College and University Environmental Programs as a Policy Problem (Part 1): Integrating Knowledge, Education, and Action for a Better World?" *Environmental Management* 47 (5): 701–15. https://doi.org/10.1007/s00267-011-9619-2.

Clark, Susan G., Murray B. Rutherford, Matthew R. Auer, David N. Cherney, Richard L. Wallace, David J. Mattson, Douglas A. Clark, Lee Foote, Naomi Krogman, Peter Wilshusen, and Toddi Steelman. 2011b. "College and University Environmental Programs as a Policy Problem (Part 2): Strategies for Improvement." *Environmental Management* 47 (5): 716–26. https://doi.org/10.1007/s00267-011-9635-2.

Ehrenfeld, John R., and Andrew J. Hoffman. 2013. *Flourishing: A Frank Conversation about Sustainability*. Stanford, CA: Stanford University Press.

Funtowicz, Silvio O., and Jerome R. Ravetz. 1995. "Science for the Post Normal Age." In *Perspectives on Ecological Integrity*, edited by Laura Westra and John Lemons, 146–61. Dordrecht: Springer Netherlands.

Geck, Caroline. 2007. "The Generation Z Connection: Teaching Information Literacy to the Newest Net Generation." In *Toward a 21st-Century School Library Media Program*, edited by Esther Rosenfeld and David V. Loertscher, 235–41. Lanham, MD: The Scarecrow Press, Inc.

King, Martin Luther, Jr. 1963. "I Have a Dream." Speech, Lincoln Memorial, Washington, DC, August 28, 1963. *American Rhetoric*. http://www.americanrhetoric.com/speeches/mlkihaveadream.htm.

Kricsfalusy, Vladimir, Colleen George, and Maureen G. Reed. 2018. "Integrating Problem-and Project-Based Learning Opportunities: Assessing Outcomes of a Field Course in Environment and Sustainability." *Environmental Education Research* 24 (4): 593–610. https://doi.org/10.1080/13504622.2016.1269874.

Lang, Daniel J., Arnim Wiek, Matthias Bergmann, Michael Stauffacher, Pim Martens, Peter Moll, Mark Swilling, and Christopher J. Thomas. 2012. "Transdisciplinary Research in Sustainability Science: Practice, Principles, and Challenges." *Sustainability Science* 7 (1): 25–43. https://doi.org/10.1007/s11625-011-0149-x.

Lasswell, Harold D., and Myres S. McDougal. 1992. *Jurisprudence for a Free Society: Studies in Law, Science, and Policy*. New Haven, CT: New Haven Press.

Masten, Ann S., J.J. Cutuli, Janette E. Herbers, and Marie-Gabrielle J. Reed. 2009. "Resilience in Development." In *The Oxford Handbook of Positive Psychology*, 2nd ed., edited by Shane J. Lopez and C.R. Snyder, 117–31. Oxford and New York: Oxford University Press.

Miller, Thaddeus R., Arnim Wiek, Daniel Sarewitz, John Robinson, Lennart Olsson, David Kriebel, and Derk Loorbach. 2014. "The Future of Sustainability Science: A Solutions-Oriented Research Agenda." *Sustainability Science* 9 (2): 239–46. https://doi.org/10.1007/s11625-013-0224-6.

Rittel, Horst W.J., and Melvin M. Webber. 1973. "Dilemmas in a General Theory of Planning." *Policy Sciences* 4 (2): 155–69.

Segran, Elizabeth. 2016. "Your Guide to Generation Z: The Frugal, Brand-Wary, Determined Anti-Millennials." *Fast Company*. September 8, 2016. https://www.fastcompany.com/3062475/most-creative-people/your-guide-to-generation-z-the-frugal-brand-wary-determined-anti-millen.

Steelman, Toddi, Elizabeth Guthrie Nichols, April James, Lori Bradford, Liesel Ebersöhn, Vanessa Scherman, Funke Omidire, David N. Bunn, Wayne Twine, and Melissa R. McHale. 2015. "Practicing the Science of Sustainability: The Challenges of Transdisciplinarity in a Developing World Context." *Sustainability Science* 10 (4): 581–99. https://doi.org/10.1007/s11625-015-0334-4.

Tett, Gillian. 2016. "Why We No Longer Trust the Experts." *Financial Times*. July 1, 2016. https://www.ft.com/content/24035fc2-3e45-11e6-9f2c-36b487ebd80a.

Williams, Alex. 2015. "Move Over, Millennials, Here Comes Generation Z." *The New York Times*. September 18, 2015. http://www.nytimes.com/2015/09/20/fashion/move-over-millennials-here-comes-generation-z.html.

11

Education for Regeneration

PATRICIA E. (ELLIE) PERKINS

IN REFLECTING on education and visions for a sustainable human society, I find inspirational this story by Michi Saagiig Nishnaabeg scholar Leanne Betasamosake Simpson, a theorist, communicator, and leader of Indigenous resurgence.[1]

> Ethically, it is my emphatic belief that the land, reflected in Nishnaabeg thought and philosophy, compels us towards resurgence in virtually every aspect. Walking through the bush last spring with my children, the visual landscape reminded me of this. We saw Lady Slippers, and I was reminded of our name for the flower and the story that goes with it,[2] and then moss, and then butterflies.[3] Then we saw a woodpecker[4] and I thought of a similar story. Finally, we walked through a birch stand and I thought of Nanabush, Niimkiig and birch bark.[5] Our Nishnaabeg landscape flourishes with our stories of resistance and resurgence, yet through colonial eyes, the stories are interpreted as quaint anecdotes with "rules" of engagement and consequence. Interpreted within our cultural web of non-authoritarian leadership, non-hierarchical ways of being, non-interference and non-essentialism, the stories explain the resistance of my Ancestors and the seeds of resurgence they so carefully saved and planted. So I could then assume my responsibility as a Michi Saagiig Nishnaabeg to care

take of their garden, eventually passing those responsibilities on to my grandchildren. (Simpson 2011, 18)

Unless readers are familiar with the other stories embedded here, reading this account leads (via the original footnotes) to tangents of new discoveries and overlapping ideas—somewhat like reading a journal article full of references, only much more multidimensional. How satisfying and simple, but erudite, it would be if readers could share this multi-layered context and background understanding of land and resurgence!

The expanding inequities and cruel power systems some humans have created—exemplified by colonialism, patriarchy, and capitalism—are now threatening all life on earth. Following centuries of imperialism, industrial "development," toxic pollution, and carbon emissions from burning fossil fuels, the status quo is not something to be sustained; rather, it must be remediated and rejected. Impacts on the most vulnerable, and the more-than-human, affect us all. Regeneration is the process of regrowing what is lost so that original functions are restored. Building a future for humans requires regenerating the Earth for all life (see Mitchell 2020). This is not the same as sustainability—a capacity to continue, endure, or be preserved over time. Regeneration expresses the conviction that continuance of catastrophic decline is not acceptable.

I write as a settler whose ancestors arrived in North America/Turtle Island hundreds of years ago, and I work within education systems that help to perpetuate these pernicious, terribly inequitable economic and social practices. This knowledge brings with it a big responsibility to learn about and support those who are changing these systems. In this chapter, the words "we," "our," and "us" refer to all who seek to contribute to regeneration by transforming education systems. Whether this is possible remains to be seen, but I believe we have a moral and intellectual responsibility to try. Building the coalitions and alliances to do so, both within and outside of education institutions, requires a diversity of voices and experiences, and it also requires deeply understanding and addressing the past and ongoing violence of colonialism and patriarchy. When we look around the classroom or meeting room, and examine the books and ideas we teach, who is missing? What voices are we not able to hear? Why are they not present, despite efforts (perhaps) to bring together a diverse group? How can those others be included?

One way is to bring in ideas transmitted in stories, or through videos or art, that would otherwise be unrepresented. Even better is to address the reasons why some cannot be present in person and are effectively excluded from the dominant education and political structures.

An emphasis on inclusivity and equity is tied to recognizing the heinous impacts of capitalist, growth-driven economic systems. We need to admit the wrong-headedness, and the impossibility, of privatizing all that is valuable. Instead, we have to learn, with humility, how to equitably respect and live within Earth's life-support systems (water, soil, air, forests, and culture) that sustain humans and all life. This requires building educational and social processes that will be capable of transmitting skills for personal and collective responsibility, conflict resolution, "two-eyed seeing,"[6] awareness of nature and others, and collaborative, appropriate behaviours. It also requires continually articulating, publicly, that individual greed is not deserving of respect or adulation—quite the opposite. Linking personal wealth with political power, as almost always happens today, is not the only or best way to run human systems; rather, it is the source of political instability. Humans can do better. An example of a way to culturally embed the redistribution of wealth and counter-balance material wealth against respect for long-term leadership (rather than allowing wealth and political/economic leadership to reinforce each other) is the traditional potlatch ceremony of First Nations on the Pacific Coast (Native American Netroots 2010; Hegmon 2017, 214). Until it was outlawed by the Canadian government in 1884, prosperous leaders publicly gave away most of their surplus to others in regular potlatch feasts, which also provided the occasion for sharing ecological knowledge, recognizing all community members, and allocating leadership positions. Says Indigenous ecological economist Ronald Trosper, "the potlatch ceremonies unified three characteristics of the Pacific Northwest system: their system of distributing wealth, their territorial system, and their governance practices" (2009, 58). He also points out: "The societies on the Northwest Coast of North America exhibited great resilience over 2000 years: how did the humans organize themselves and their neighbours into a social-ecological system that sustained itself?...Ecological economists and other students of sustainability need to examine examples of societies that have a track record of resilience" (2009, ix; see also Wilner 2013 and Umeek 2011).

Lived Experience, Diversity, and Systems

Rather than training people to expect privilege through credentials (accepting the valuation scheme and normative values of capitalism), how can education help people learn to centre land, in all its wondrous complexity, and live within ecological systems? The beginning steps toward educational transformation, as many others have noted, include elements such as implementing experiential education, incorporating team building and collaboration, spending time outdoors, ensuring basic science and systems literacy, practicing with arts and multiple ways of knowing, ensuring transdisciplinarity, commoning (building collective rather than individual relationships with life necessities), removing perverse academic silos, and making ongoing efforts to decolonize and de-Westernize communication and thought (Orr 2004; Orr 2016; Stone and Barlow 2005; Berkes 2017; Molnár et al. 2019; Styres 2017; Tuck, McKenzie, and McCoy 2014; McCoy, Tuck, and McKenzie 2016; Henry 2014; Smith, Tuck, and Yang 2018). Of these, systems literacy, commoning, and experiential education seem particularly salient to me, because they help articulate and prefigure heightened ecological understandings of human well-being and humanity's place in the world in relation to other beings. They also help build alternatives to colonialism, patriarchy, and capitalism, so they are steps in the direction of transformation and resurgence.

Systems literacy, like language or mathematical literacy, is a doorway to understanding the world's complexities and interrelationships. Systems cannot be well described using binaries, rigid classification, reductionist modelling, and other basics of Western colonial worldviews. Systems analysis also reveals the deep error of allowing capitalist commodification or extraction of one piece of a system. Teachers and mentors need to help students develop a kind of systems-literate common sense—the ability to distinguish between things that make no sense at all, those that seem sensible but that are also problematic on other levels, and those that really lead "in a good direction."[7] The best pedagogical tools are collaborative lived experience through challenging field trips, outdoor education, sometimes-difficult group work, and gradual, layered discovery of socio-ecological systems in order to explore their complex feedbacks, intricacies, and surprises, which impart awe and humility (Berkes, Colding, and Folke 2003; Krasny, Lundholm, and Plummer 2011).

One of the metaphorical frames that I find useful for thinking and teaching about systems is the idea of watersheds. A watershed is a system in which the energy that enters via wind and rainfall is virtually limitless, while in a material sense the flows are more constrained. Fuelled by solar energy, water rains down at the top of the system and washes soil nutrients (and any wastes or pollutants) downhill, affecting the lives and interactions of the people and other living and non-living things farther down within the watershed. The relationships between people who live on the heights and in the lowlands are interconnected, ecologically and socially. Thinking about where we are in the watershed and how our actions affect others and interact with all else that happens is a way of linking our thoughts to the land and to other people, thus helping to build a collective politics.

The attentiveness and common sense that I believe most people share is undermined by simplistic, non-empirical leadership and decision making that strains common sense (such as subsidies that support fossil fuels) and by the way that our attention is frittered away with constant demands to multi-task. An example is that, even in class, many students are likely to be on Facebook, rather than paying attention to the topic at hand. Somehow, as teachers, we have to be able to name this as a political issue, turn it around, and refocus collective attention. We can point out that people have many interests; they have lived experiences in many different places and valuable knowledge to share, so how can we bring everyone together and build on that diversity to shape our collective present and future? How can we use the skills and the connections that all in the room have, to think together about what needs to happen?

In my view, the crisis we are now living in is related to many people not being able to envision a way toward replacing globalized capitalism, which is driving the world toward environmental disaster, with another kind of aware, equitable, collective politics that can lead to the regeneration of our home, the Earth. This is linked to the crisis of making the energy transition beyond fossil fuels. These crises overlap, but they are not exactly the same. They share aspects of fear, denial, guilt, and shame—all negative emotions on the part of those of us who know we consume too much and are responsible for the worst impacts of the crises, and that we must therefore try to "turn the canoe around" or take action to change the situation for the better (Klein 2013; see also Cole 2006). But how? We must figure it out together.

Participatory classrooms and opportunities to explore complex systems outside the classroom, with teachers who model how to create safe spaces for discussion in diverse groups and welcome participation from all, are important pillars of education through lived experience with biodiversity.

Commoning and Equity

As teachers, we also have a responsibility to show some glimmers of hope and possible ways of moving forward to resolve these crises. One of those glimmers for me was when Elinor Ostrom was awarded the Nobel Prize in Economic Sciences in 2009 for her studies on the conditions under which people can develop sustainable governance systems that prevent open access to the "common-pool resources" used by many, thus preventing the "tragedy of the commons." The community attributes for successful commons governance that Ostrom (1990) identified in her research include features like mutual knowledge and respect, a bounded system so that people recognize the limits of the "resource," a history of regulations developed in a participatory way with enforcement so that people know that norms cannot be violated with impunity, and non-interference by higher orders of government in the local community's own governance system. Ostrom (2010a, 2010b, 2014) also developed the systems-relevant concept of "polycentricity," which explains how different levels of authority and different kinds of skills can interact with each other to make governance systems work better and with more resilience. She showed that a polycentric system is not inefficient, even though it has overlapping functions; instead, it is stronger and more sustainable.

These are ideas that fly in the face of, and that contradict, many of the basic "efficiency" tenets of economics, which after all is a description of life under capitalism. In my view, and in the view of a growing number of scholars who are working on climate change and climate justice, among other topics, this commoning approach is very hopeful. It revives and underscores the importance of participatory democracy and local responsibility for standing up to private interests and preventing the commodification of water, mineral resources, forests, fisheries, etc. Commons governance is fundamentally different from allowing markets to run things. It is also different from centralized state control and planning. It is different from the kind of hybrid system that now exists in most places, where governments intervene in market-based economic systems to nudge them in various directions, usually designed to help the interests of the powerful. "Free markets" have never been a realistic description of how political economy really operates anyway, as feminist ecological economists have documented; unpaid care work and "free" inputs from "nature," made possible by control over women, marginalized peoples, and common "resources," have always undergirded capitalist economies (Mies 1986; Mellor 1992; Salleh 2009). The market economy is just the tip of the iceberg; it is supported by unpaid work, natural systems, and ecosystem services, all of which are much larger than the economy that we are trained to see. There is a deep alliance between feminist and ecological critiques of capitalism, and a widening recognition that patriarchy, colonialism, and capitalism are not suited or acceptable as ideological frames for resolving current crises since they are causally implicated (Folbre 2021; Santos 2018; Fortier 2017).

Cooperatives and commons are more prevalent and more important in assuring people's livelihoods globally than many may realize. The United Nations has estimated that the livelihoods of half the world's population are made secure by cooperative enterprises (COPAC 1999, 1). Mutual aid, utopian communities, grassroots collaborative economic initiatives, and co-ops allowed Black Americans to persevere in "finding alternative economic strategies to promote economic stability and economic independence in the face of fierce competition, racial discrimination, and White supremacist violence and sabotage," while building leadership and community stability (Gordon Nembhard 2014, 28). Access to community-managed resources and services still undergirds resilience by helping people all over the world to weather personal, economic, and ecological shocks (Brown 2015, 156). Forward-looking education systems must teach these histories as well as the skills for strengthening and rebuilding commons: respect, creative institution building, dispute resolution, intolerance for injustice, and cultural awareness.

Regeneration

I believe there are cracks in the current unsustainable, crisis-ridden, political and economic systems; through those cracks grows awareness of the importance of alternative livelihood systems like commons and how humans can build and transmit the collective skills to regenerate and preserve them. Dene activist Glen Coulthard, in his book *Red Skin White Masks*, speaks about this hope and the promise of commons.

> What must be recognized by those inclined to advocate a blanket "return to the commons" as a redistributive counterstrategy to the neoliberal state's new round of enclosures, is that, in liberal settler states such as Canada, the "commons" not only belong to somebody—*the First Peoples of this land*—they also deeply inform and sustain Indigenous modes of thought and behavior that harbor profound insights into the maintenance of relationships within and between human beings and the natural world built on principles of reciprocity, nonexploitation and respectful coexistence. By ignoring or downplaying the injustice of colonial dispossession, critical theory and left political strategy not only risks becoming complicit in the very structures and processes of domination that it ought to oppose, but it also risks overlooking what could prove to be invaluable glimpses into the ethical practices and preconditions required for the construction of a more just and sustainable world order. (2014, 12; emphasis original)

To me, Coulthard is pointing out that it would be a grave and futile error for others to just take Indigenous ideas and try to apply them within colonial systems. By "colonial," I mean industrial, fossil fuel-based capitalist and patriarchal economies, water-based sanitation, and the idea that everyone has a right to a personal metal transportation pod that spews carbon into the air. I mean plastic, fast fashion, vast diversities of consumption goods, along with related ideas and "rights" that have become normalized for many over the past one hundred years or so, the absence of which might be scary to those of us who have enjoyed these things for our whole lives, and whose parents may have, too, but whose grandparents probably did not; these are aberrations in human history. Humans can create better, healthy, durable, and equitable ways of living on the Earth. We need to help each other in order to see how change is positive, not just fear inducing, and to learn how to carry out change processes while protecting the most vulnerable.

To do that, we have to get students out of the classroom, both literally and epistemically, so that they are engaged with the natural life-world and with people beyond academia. This will get them involved with the kind of sensitive, equity-enhancing problem solving that is vital.

We should also take advantage of students' interests and strengths with social media, networking, and online communication across difference and across the globe. The biggest problem that I see in young people now is boredom, anomie, and the sense that it does not matter what any one person does because power structures are entrenched and things will not really change. This lack of hope is a difficult problem. I see young people's networking skills as a potential way to address it, collectively.

Conflict resolution, systems literacy, and the importance of diversity, colonial history, commoning, and decolonization need to be part of basic education, not just in sustainability programs but throughout universities and for adults, too, in the broader society. We have to model creative, activist lifestyles, and create governance systems that include the views, priorities, and needs of the vulnerable, thus removing much of the fear that accompanies transitions. We need to recognize and acknowledge commons where they are existing and emerging, and remove barriers to commoning through research, pedagogy, and practice. The connections between pedagogy, activism, and research are fundamental; they all reinforce each other.

Conclusion

I gain so much energy and inspiration from witnessing the accomplishments of students and former students I have been fortunate to know. When things need to be done, young people create NGOs and community organizations. They build community and make videos and art, and their social action knows no bounds. Young people have been the mainstays of Idle No More, Occupy, Fridays for the Future, Black Lives Matter, and many other campaigns fuelled by social networking. The legacy of these movements is their educational record among those whose lived experience motivates their will to continue to build participatory social change.

Commons-building organizations started by students and other young people are everywhere. They include, among many, many more:

- Not Far From the Tree (http://notfarfromthetree.org/)
- Centre for Social Innovation (https://socialinnovation.org/)
- Great Lakes Commons (http://www.greatlakescommons.org/)
- Jane Finch Action Against Poverty (https://jfaap.wordpress.com/)
- Mining Injustice Solidarity Network (https://mininginjustice.org/)
- Indigenous Climate Action (https://www.indigenousclimateaction.com/)

I recommend taking time to reflect on the creative contributions of young people we have known, as a way of facing the future with hope and conviction that diverse communities of human beings can collectively regenerate the Earth, starting with education, provisioning, and governance—indeed, we are already doing so.

Notes

1. Simpson (2016, 22) says, "Indigenous resurgence, in its most radical form, is nation building, not nation-state building, but nation building, again, in the context of grounded normativity, by centring, amplifying, animating, and actualizing the processes of grounded normativity as flight paths or fugitive escapes from the violences of settler colonialism." She quotes Glen Sean Coulthard's definition of "grounded normativity": "the systems of ethics that are continuously generated by a relationship with a particular place, with land, through the Indigenous processes and knowledges that make up Indigenous life" (Coulthard 2014, 60). See also Simpson (2011, 17).
2. "For a written version of this story, see Lise Lunge-Larsen and Margi Preus, *The Legend of the Lady Slipper*, Houghton Mifflin, 1999." (Simpson 2011,28)
3. "One version of this story exists in 'The First Butterflies' in *Tales the Elders Told: Ojibway Legends* by Basil Johnston, Royal Ontario Museum, Toronto ON, 1983, 12-17; another exists in John Borrows' *Drawing Out Law: A Spirit's Guide*, University of Toronto Press, Toronto ON, 2010, 14-16." (Simpson 2011, 28)
4. "Basil Johnston, 'The Woodpecker' in *The Bear-Walker and Other Stories*, Royal Ontario Museum, Toronto ON, 1983, 49-55." (Simpson 2011, 28)
5. "*Niimkiig* means thunderbirds. For a version of this story, see Wendy Makoons Geniusz's 'Nenabozho and the Animkikiig' in *Our Knowledge is Not Primitive:*

Decolonizing Botanical Anishinaabe Teachings, Syracuse University Press, Syracuse NY, 2009, 136–40." (Simpson 2011, 28)

6. As discussed at the forum of all the contributors to this book at the Banff Conference Centre, September 21–23, 2016, this means seeing with both a Western eye and an Indigenous eye, engaging with diverse "ways of knowing," including ethical and cultural traditions and recognizing the appropriateness of knowledge that is grounded in diverse ways.

7. The Nishnaabemowin term *mino bimaadiziwin* means "in a good way" or "continuous rebirth." See Simpson (2011, 26).

References

Berkes, Fikret. 2017. "Environmental Governance for the Anthropocene? Social-Ecological Systems, Resilience, and Collaborative Learning." *Sustainability* 9, no. 7 (July): 1232. https://doi.org/10.3390/su9071232.

Berkes, Fikret, Johan Colding, and Carl Folke, eds. 2003. *Navigating Social-Ecological Systems: Building Resilience for Complexity and Change.* New York: Cambridge University Press.

Borrows, John. 2010. *Drawing Out Law: A Spirit's Guide.* Toronto: University of Toronto Press.

Brown, Katrina. 2015. *Resilience, Development and Global Change.* New York and London: Routledge.

Cole, Peter. 2006. *Coyote and Raven Go Canoeing: Coming Home to the Village.* Montreal: McGill-Queen's University Press.

COPAC (Committee for the Promotion and Advancement of Cooperatives). 1999. "The Contribution of Cooperatives to the Implementation of the World Summit for Social Development Declaration and Programme of Action." Conference Room Paper for the First Session of the Preparatory Committee for the Special Session of the General Assembly on the Implementation of the Outcome of the World Summit for Social Development and Further Initiatives, New York, 17–28 May 1999. Geneva: COPAC. http://www.copac.coop/publications/1999-coops-wssd5.pdf.

Coulthard, Glen Sean. 2014. *Red Skin White Masks: Rejecting the Colonial Politics of Recognition.* Minneapolis/London: University of Minnesota Press.

Folbre, Nancy. 2021. *The Rise and Decline of Patriarchal Systems.* New York: Penguin Random House.

Fortier, Craig. 2017. *Unsettling the Commons: Social Movements within, against, and beyond Settler Colonialism.* Winnipeg: ARP Books.

Geniusz, Wendy Makoons. 2009. *Our Knowledge is Not Primitive: Decolonizing Botanical Anishinaabe Teachings.* Syracuse: Syracuse University Press.

Gordon Nembhard, Jessica. 2014. *Collective Courage: A History of African American Cooperative Economic Thought and Practice.* University Park: Pennsylvania State University Press.

Hegmon, Michelle, ed. 2017. *The Give and Take of Sustainability: Archaeological and Anthropological Perspectives on Tradeoffs.* New York: Cambridge University Press.

Henry, Elizabeth. 2014 "A Search for Decolonizing Place-Based Pedagogies: An Exploration of Unheard Histories in Kitsilano Vancouver, B.C." *Canadian Journal of Environmental Education* 19: 18–30.

Johnston, Basil. 1981. *Tales the Elders Told: Ojibway Legends*. Toronto: Royal Ontario Museum.

Johnston, Basil. 1995. *The Bear-Walker and Other Stories*. Toronto: Royal Ontario Museum.

Klein, Naomi. 2013. "Embodying the Transformation of Idle No More: In Conversation with Leanne Simpson." *Rabble.ca*, March 6, 2013. http://rabble.ca/columnists/2013/03/embodying-transformation-idle-no-more-conversation-leanne-simpson.

Krasny, Marianne E., Cecilia Lundholm, and Ryan Plummer, eds. 2011. *Resilience in Social-Ecological Systems: The Role of Learning and Education*. New York and London: Routledge.

Lunge-Larsen, Lise, and Margi Preus. 1999. *The Legend of the Lady Slipper*. Boston: Houghton Mifflin.

McCoy, Kate, Eve Tuck, and Marcia McKenzie, eds. 2016. *Land Education: Rethinking Pedagogies of Place from Indigenous, Postcolonial, and Decolonizing Perspectives*. London and New York: Routledge.

Mellor, Mary. 1992. *Breaking the Boundaries: Towards a Feminist Green Socialism*. London: Virago Press.

Mies, Maria. 1986. *Patriarchy and Accumulation on a World Scale: Women in the International Division of Labour*. London and Atlantic Heights, NJ: Zed Books.

Mitchell, Audra. 2020. "Revitalizing Laws, (Re)-making Treaties, Dismantling Violence: Indigenous Resurgence against 'the Sixth Mass Extinction.'" *Social & Cultural Geography* 21, no. 7 (October): 909–24. https://doi.org/10.1080/14649365.2018.1528628.

Molnár, Zsolt, Leticia Doormann, Victoria Reyes-García, Berta Martin-Lopez, Fikret Berkes, and Örjan Bodin. 2019. "Learning from Indigenous Populations and Local Communities." *One Earth* 1, no. 1 (September): 16–17. https://doi.org/10.1016/j.oneear.2019.07.002.

Native American Netroots. 2010. "The Potlatch." August 13, 2010. http://nativeamericannetroots.net/diary/631.

Orr, David W. 2004. *Earth in Mind: On Education, Environment, and the Human Prospect*. 10th Anniversary Edition. Washington, DC, Covelo, CA, and London: Island Press.

Orr, David W. 2016. *Dangerous Years: Climate Change, the Long Emergency, and the Way Forward*. New Haven and London: Yale University Press.

Ostrom, Elinor. 1990. *Governing the Commons: The Evolution of Institutions for Collective Action*. New York: Cambridge University Press.

Ostrom, Elinor. 2010a. "Beyond Markets and States: Polycentric Governance of Complex Economic Systems." *American Economic Review* 100, no. 3 (June): 641–72. https://doi.org/10.1257/aer.100.3.641.

Ostrom, Elinor. 2010b. "Polycentric Systems for Coping with Collective Action and Global Environmental Change." *Global Environmental Change* 20, no. 4 (October): 550–57. https://doi.org/10.1016/j.gloenvcha.2010.07.004.

Ostrom, Elinor. 2014. "A Polycentric Approach for Coping with Climate Change." *Annals of Economics and Finance* 15 (1): 97–134.

Salleh, Ariel, ed. 2009. *Eco-Sufficiency and Global Justice: Women Write Political Ecology*. London: Pluto Press.

Santos, Boaventura de Sousa. 2018. *The End of the Cognitive Empire: The Coming of Age of Epistemologies of the South*. Durham, NC: Duke University Press.

Simpson, Leanne Betasamosake. 2011. *Dancing on Our Turtle's Back: Stories of Nishnaabeg Re-creation, Resurgence and a New Emergence*. Winnipeg: ARP Books.

Simpson, Leanne Betasamosake. 2016. "Indigenous Resurgence and Co-resistance." *Critical Ethnic Studies* 2, no. 2 (Fall): 19–34. https://doi.org/10.5749/jcritethnstud.2.2.0019.

Smith, Linda Tuhiwai, Eve Tuck, and K. Wayne Yang, eds. 2018. *Indigenous and Decolonizing Studies in Education: Mapping the Long View*. New York and London: Routledge.

Stone, Michael K., and Zenobia Barlow. 2005. *Ecological Literacy: Educating Our Children for a Sustainable World*. San Francisco: Sierra Club Books.

Styres, Sandra D. 2017. *Pathways for Remembering and Recognizing Indigenous Thought in Education: Philosophies of Iethi'nihsténha Ohwentsia'kékha (Land)*. Toronto, Buffalo, and London: University of Toronto Press.

Trosper, Ronald L. 2009. *Resilience, Reciprocity and Ecological Economics: Northwest Coast Sustainability*. New York and London: Routledge.

Tuck, Eve, Marcia McKenzie, and Kate McCoy. 2014. "Land Education: Indigenous, Post-Colonial, and Decolonizing Perspectives on Place and Environmental Education Research." *Environmental Education Research* 20 (1): 1–23. https://doi.org/10.1080/13504622.2013.877708.

Umeek (E. Richard Atleo). 2011. *Principles of Tsawalk: An Indigenous Approach to Global Crisis*. Vancouver: UBC Press.

Wilner, Isaiah. 2013. "A Global Potlatch: Identifying the Indigenous Influence on Western Thought." *American Indian Culture and Research Journal* 37 (2): 87–114. https://doi.org/10.17953/aicr.37.2.70431618hr053470.

V | Unique Perspectives from Professor and Student

12

Education for Sustainability
An Ecological Citizenship Approach in a Neoliberal Age

ALLISON F.W. GOEBEL

I HAVE BEEN CO-TEACHING our introductory course in environmental studies at Queen's University from its origin in 2003.[1] I have also participated in several curriculum reviews and developments for both our undergraduate and graduate programs and been involved in the supervision of many graduate students in environmental studies. Our program has been graduating top-level students from our various honours programs since the late 1990s, sending them off into the world, hopefully to make a positive impact on environmental issues. Over this same period, most major environmental issues—such as climate change, water issues, biodiversity loss, and the spread of chemical toxins—have only gotten worse. In March 2016, the Canadian icon of environmentalism, David Suzuki, reflected on his eightieth birthday that the environmental movement had failed (CBC News 2016). He declared that we have not made the necessary changes to address environmental catastrophe. Indeed, in the five or six years since then, mainstream attention has grown increasingly alarmist, most notably with the worldwide climate strikes led by Swedish teen Greta Thunberg (Irfan 2019).

    The globe as a whole has been shaken by evident economic fragility, especially since the 2008 financial crash and then the global economic downturn triggered by the COVID-19 pandemic in 2020. As well, social and political instability is signalled by growing inequalities within and between nations, mass refugee crises, and civil wars. The election of Donald Trump for president of the United States in 2016, a vocal climate change denier, added to the gloom. His presidency saw attacks on world governance institutions such as the United Nations, growing hostility toward China, and massive anti-racism protests and race riots. As one participant at the Banff Forum[2] put it, we are living in an age of "apocalyptic disappointment."

At the same time, the student body has changed profoundly in my more than twenty years of university teaching, not only as a result of social media and other aspects of the information age, but as it is a generation deeply saturated in intensified individualism created through deepened neoliberalism. As students have retreated into the algorithm-designed individualized worlds of social media platforms, they appear distracted in class and easily bored. In lecture classes, it is common for more than half of the students to not even attend. Partly in response to this, we are called upon to change our teaching techniques, especially by incorporating online learning tools and innovating beyond the "sage on the stage" lecture-style mainstay of university teaching. Experiential and inquiry-based learning are the colours of the day.

On a personal level, I have lost all of my elders (father, mother, and uncle) and found myself in the disorienting position of head of my extended family. I have become middle aged, but could I ever be an elder? I feel the profound need to be a better teacher and leader, and a deepened sense of responsibility to prepare my students for the world.

In this chapter, I want to reflect on our approach to teaching our introductory course and suggest it is time for something new that directly addresses the changing social, environmental, and political context and more thoughtfully reflects our roles as teachers and leaders for positive changes toward sustainability. For the purposes of this chapter, my working definition of sustainability rests upon a broadly conceived philosophy about a process that promotes human and natural systems "thriving"—creating more abundance and enrichment where the need and opportunity for change converge (Edwards 2010). Our team won a Principal's Dream Course competition at Queen's, which gave us the resources to remodel the course around inquiry-based learning approaches and to generally shake up the course.[3] More on inquiry-based learning below.

Early Approaches

In teaching our introductory course, we have always tried to avoid the "doom and gloom" message of environmental catastrophe, conscious of how this depresses and disempowers students and can lead to disengagement and even despair (McKinley 2008). While not avoiding the details and seriousness of environmental problems, we have also focused on strategies for change, such as research and activism, and attempted to offer glimmers of hope and success (don't we all trot out the Montreal Protocol as a global agreement success story?). We have placed a lot of focus on practical tools, such as sustainability frameworks, interdisciplinarity, team research concepts and practices, critical thinking, and writing and presentation skills. We have used exercises that practice engagement, such as writing to politicians, critically analyzing media stories, and completing research projects on local environmental problems. We have used tools of self-measurement, such as ecological footprint and water footprint exercises, which of course always turn out badly for the students and promote guilt and shame. One thing this has led me to consider is that we do not necessarily know how to help students productively manage this guilt and shame. Is this part of our pedagogical and ethical role? We also know little about the nature of students' emotions beyond this surface observation of guilt and shame. Are students experiencing the broader, more complex existential despair that many in our own generation have started to express, the "spectral haunting that comes from more-than-human loss" (Cunsolo and Landman 2017, 3), or in other words, the emotional impacts of large-scale ecological loss (species extinction, contamination of loved places, loss of traditional territories, etc.)? I pursue this thought further below.

In one way we have definitely succeeded: we (including environmental studies as a whole in North America) have become very good at studying and explaining environmental issues, including their scientific, social, and political aspects. Where we have found less success is in figuring out how to solve these problems, especially in relation to the limitations of individual action. Yes, I can reduce my water and energy use, buy "greener" products, and ride my bicycle, but the suburbs continue to encroach on our best farmland, the climate is still changing, and more developed countries are still dumping toxic electronic waste on the poor in West Africa and Asia. While we talk about the need for broader political engagement and introduce the idea of ecological citizenship (Dobson 2003, 2007), I think we have not done enough to equip students to effectively participate in the social, institutional, and structural changes that are necessary for sustainability, given the powerful counter forces of neoliberal individualism and the imperatives for growth of global capitalism.

Next Step: Ecological Citizenship?

The rollback of the state and enhanced focus on individual responsibility as a mode of citizenship has characterized what has come to be known in the politics of Western democracies as neoliberalism, which is usually described as beginning in the 1980s under Ronald Reagan (US) and Margaret Thatcher (UK). This prevailing political climate—which deeply affected Canada, especially under the leadership of former prime minister Stephen Harper—is fertile ground for a kind of mainstream environmentalism that promotes individual responsibility for environmental action. In this paradigm, it is up to individual citizens to choose to reduce their environmental impact through recycling, reducing their own energy and other consumption, and demanding "green" products as consumers (Cao 2015, 63). This paradigm also holds that governments should be less involved in protecting public goods (such as education, social welfare, and environmental health), and that they should reduce constraints on private business. This is in the belief that corporations will perform better, create more jobs, and be motivated to self-regulate negative environmental and social impacts if that is what their customers demand.

This is the political context in which our students have come of age, and they respond readily to calls to "green" their consumption, but not as well to critiques of the growth paradigm of global capitalism. Partly, this has to do with the fact that managing their own consumption is at least somewhat in their control. Reigning in the forces of corporate power seems way beyond their powers.

In revamping our course, then, a central theme was to challenge individualism and experience the limits of individual action. Using experiential and inquiry-based learning, we guide students through an experience that hopefully leads them to an appreciation of the need to foster ecological citizenship, build social relations and institutions, and challenge the ideological and material tenets of neoliberalism that leave environmentalism to the voluntary actions of individuals and corporations. In chapter 1 of this volume, Bergstrom discusses the tenets of experiential and problem-based learning and mentions key scholars such as Domask (2007) and Wiek et al. (2014), who speak to the potential for courses like mine to foster self-directed learning that deepens students' critical understanding of real-world sustainability problems.

By lucky coincidence, my son, Gabes Epprecht, was doing his honours thesis in environmental studies at Trent University in 2015–16. He chose to do an autoethnography of his experience of trying to eat outside of the industrial food system, incorporating gardening, foraging wild foods, and buying only local food. In the process, he confronted his own rampant individualism and arrogance, and ultimately his powerlessness, shame, and despair in the face of the inescapability of the industrial food system. Our teaching team decided to use Gabes' work and invited him to participate in our class as a person only a few years older than our students, someone much closer to their lived realities. We incorporated excerpts from Gabes' thesis and recorded video clips that tracked his transition from focusing on individual action to acknowledging the critical need to build social relations and community to achieve environmental transformation. A few excerpts from his thesis are compiled in the following:

> I began this project with a sincere desire to reduce my environmental impact and change my relationship with food and the natural world. Being a confident, and somewhat competent person, I decided that the best way for me to do this was to completely change the way I ate and actively resist the industrial diet through personal initiatives. Overwhelmed by the scale and forms of environmental degradation, and at a time of personal re-identification, I felt the need to latch onto tangible forms of environmental resistance and prove to myself that I was capable of a pro-environmental transition…Following my numerous failed attempts at personal dietary change, I fell into a state of depression. I began to question my self-worth, rage against the fact of anthropogenic environmental degradation and belittle industrial

society from the desperate position of a crumbling moral pedestal... I was not roused from this bleak mindset until I began to understand the value of creating positive social interactions, and accept the important role that larger physical, social, and economic structures played in determining my diet...*Makes me feel small, meaningless. And really angry. My powerlessness in the face of the destructive system of industrial society and my own embeddedness within* [it] *is maddening. I thought I was better! I thought cleverness and cynicism, and white male privilege was enough to separate me from guilt. From responsibility, from any sort of challenge. Shouldn't all the doors open? When I want healthy environmentally sustaining food shouldn't it just be provided to me?* (Epprecht 2016, 33–34; emphasis original)

Using a set of online modules that follow a step-by-step process of inquiry-based learning, our students, like Gabes, explore their passions to do something about an environmental issue, test their individual capacity to achieve something, and then hopefully begin to reach for a larger engagement in building social relations, institutions, and structures that challenge individualism. We hope for them to end up somewhere beyond frustration with the limitations of individual action, and to perhaps become more ready to engage with the wider world through the building of social relations, community, and public engagement.

More specifically, the modules require students to choose one environmental issue that they feel passionately about from among those discussed in class. They practice developing questions related to their topic to arrive at a focused question to pursue through other steps. They follow an inquiry-based process to explore the issue at hand, which encourages autonomous undergraduate research (Willison and O'Regan 2007), as well as their individual responsibilities and responses to the issue. This process is mostly self-directed (with help from teaching assistants), and includes a set of questions such as: What do I need to know about my issue? Where do I find reliable information about it? What are the key facts? What can I do myself about this issue? Students then have time to "do something about it" and then ask themselves: How hard or easy was it to do what I did? What barriers or incentives did I face? How big and what kind of a difference will this make? How did I feel about these activities and the learning that happened?

Students are then to link their individual explorations to the critical coursework and concepts of sustainability and knowledge about environmental issues. These connections, in turn, are meant to support students' understanding of the complexity of "doing something" to improve sustainability, which requires both individual and social engagement and transformation. The activities are also meant to engage them in an exploration of how they—as students—might engage in the wider community on campus or through non-governmental organizations, city governments, media, etc. Thus, a final set of guiding questions for this work is: What can I get involved in for broader social–institutional change toward sustainability on my issue? How hard or easy will this be? What are the barriers I face or incentives for getting involved in this way? What might I be able to achieve? What are the implications for me and my engagement with the world based on what I learned or found out? A good example of a successful process was one student's work on organic agriculture: she came to recognize that joining the local food movement (collective action) was a necessary step to take beyond simply purchasing organic food (individual action), and that she faced barriers—including social stigma, accessibility, and price—that hampered individual choice. Another student grappled with the need to reduce energy consumption and promote environmentally sustainable energy generation. In the process, she identified both micro level barriers (e.g., an argument with her roommate over the thermostat) and macro level barriers (such as government policy) and recognized the need for collective political and awareness building action.

Finally, we are addressing the feelings of shame, anger, and despair often experienced by students who confront the issues of our time and begin to understand their place in this. Addressing these emotions has to be a part of sustainability teaching, as part of pursuing human and environmental flourishing. We feel we need to do this despite (or even also because of) our own generation's emotional difficulties and our own struggles to stay positive in the face of our knowledge of the extent of the catastrophe: "We surround ourselves with the pain of the world, and try to keep ourselves well enough to move forward in this important work.

But it is not easy. In fact, it is often really damn hard to keep going" (Cunsolo 2017, xvi). In directly addressing the emotions related to sustainability learning for both faculty and students, we make ourselves as teachers more vulnerable and human and also make space for and validate the often overwhelming emotions experienced by our students.

In Fall 2016, we participated in a research study about how students in introductory courses on sustainability respond intellectually and emotionally to the course content.[4] Findings emerging from this study reveal the intensity of emotions among students (Kuyvenhoven and Graham 2017). Here are a few quotations from surveys completed by our students that express betrayal, anger, and frustration:

> It is upsetting that I only learned about so many things now. This should have been presented way earlier and there should be put more effort in raising awareness in general

> EVERYTHING made me angry. Although it was great to learn about some positive sides to sustainability that I may not have heard about, it still made me very frustrated as there is so much to do, and so many barriers that come up on the way to making change (emphasis original)

Students also expressed determination to get involved to promote change, however:

> It has made me even more determined

> I am more conscious about my actions and the consequences

> It is both depressing and inspiring to me to see what we have done and what we can do

Our own team followed up with a survey and talking circle research project with our 2017 class. One of the main objectives of this research was to further probe the emotional responses of students to sustainability learning. The survey was administered twice, first early in the term and then again near the end of the term, to measure if emotional responses changed as students learned from course content. In the first round, students expressed sixty-three different emotions in relation to a range of environmental problems, the most common being sadness, worry, concern, disgust, and disappointment. In the second round of the survey, near the end of the course, students expressed being more optimistic. The most common emotions expressed were hopeful, informed, and optimistic. This suggests that more knowledge about the issues actually helped students feel more positive. However, students also expressed being concerned, determined, disappointed, and obligated in relation to some environmental issues they had learned about in the course. We have incorporated what we learned about student emotions into subsequent iterations of the course to more deliberately address and support the emotional side of sustainability learning.[5] More specifically, we have included discussions of emotional responses to environmental issues in our introductory lectures and incorporated an optional element in the inquiry-based learning modules for students to explore and express their emotional responses to their environmental issue. We also include discussions about the emotional dimensions of sustainability learning in our training of teaching assistants, who carry out most of the face-to-face learning with the students.

Conclusion

We have been through several rounds of our revised course in environment and sustainability. Some students have hesitated at the openness of the invitation to follow their heart in choosing a topic to pursue in the inquiry-based modules, often displaying self-doubt or uncertainty. Others have jumped in with both feet and run with the opportunity for self-exploration. The emotional work of acknowledging the scale of ecological loss, our complicity in it, and the seriousness of the threat to human survival is painful and hard, for faculty as well as for students and teaching assistants. However, there is some evidence from our work that addressing the emotional side of sustainability learning supports students in moving toward a more positive, engaged outlook. Challenging neoliberalism is also hard, but we see evidence in our students' projects that they are taking on this challenge. They are learning to challenge the dominant social paradigm of neoliberal economic growth and individualism and are engaging in social relations and institution building. Students are finding places to do this work in local food movements, social justice work, and fossil fuel divestment campaigns, among other actions. The approach we are taking requires courage and no small amount of care for ourselves as teachers, but as sustainability educators, we know our subject is unlike others.

Our inquiry-based model, which recognizes the emotional side of student learning, makes small steps in the pursuit of sustainability and both human and environmental thriving. Students develop and practice self-efficacy, confront their humanness and vulnerability, and explore their connectedness to other people, institutions, and the environment around them. Moving forward, our approach could be deepened by a multi-year program that supports students in deliberately following their emotional, intellectual, and political learning beyond the first-year course, so that by their fourth year, they are prepared, as ecological citizens, to leave us, less confused, less hurt, more engaged, and ready to thrive.

Notes

1. ENSC203 Environment and Sustainability (formerly Explorations in Environmental Studies), School of Environmental Studies, Queen's University, Kingston, Ontario. This course has always been co-taught by a social scientist (usually me) and a scientist (mostly Stephen Brown, chemistry). The course was changed to a first-year offering (ENSC103) in 2017.
2. The Banff Forum was a three-day forum that took place in September 2016 in Banff, Alberta, Canada. At the forum, fourteen participants from Canada and the United States discussed the future of sustainability science and education at North American universities.
3. The team that created our winning proposal included Professor Stephen Brown (co-teacher), Professor Alice Hovorka (head, School of Environmental Studies), and myself.
4. This study, "The Experiential Dimension of Sustainability Courses," was conducted by Peter Graham and Cassandra Kuyvenhoven, PHD candidates at Queen's University, School of Environmental Studies, with the cooperation and supervision of professors Rena Upitis (Department of Education, Queen's University) and Adeela Arshad-Ayaz (Department of Education, Concordia University).
5. A paper based on this research has been published by Peter Graham et al. (2020).

References

Cao, Benito. 2015. *Environment and Citizenship*. London and New York: Routledge.

CBC News. 2016. "David Suzuki Talks Trudeau, Aging and Failed Environmentalism as He Turns 80." *CBC News*, March 16, 2016. http://www.cbc.ca/news/canada/david-suzuki-turns-80-cbc-the-national-1.3492707.

Cunsolo, Ashlee. 2017. "Prologue: She was Bereft." In *Mourning Nature: Hope at the Heart of Ecological Loss and Grief*, edited by Ashlee Cunsolo and Karen Landman, xiii–xxii. Montreal and Kingston: McGill-Queen's University Press.

Cunsolo, Ashlee, and Karen Landman. 2017. "Introduction: To Mourn beyond the Human." In *Mourning Nature: Hope at the Heart of Ecological Loss and Grief*, edited by Ashlee Cunsolo and Karen Landman, 3–26. Montreal and Kingston: McGill-Queen's University Press.

Dobson, Andrew. 2003. *Citizenship and the Environment*. Oxford: Oxford University Press.

Dobson, Andrew. 2007. "Environmental Citizenship: Towards Sustainable Development." *Sustainable Development* 15 (5): 276–85.

Domask, Joseph J. 2007. "Achieving Goals in Higher Education: An Experiential Approach to Sustainability Studies." *International Journal of Sustainability in Higher Education* 8 (1): 53–68. https://doi.org/10.1108/14676370710717599.

Edwards, Andrés R. 2010. *Thriving beyond Sustainability: Pathways to a Resilient Society*. Gabriola Island, BC: New Society Publishers.

Epprecht, Gabriel. 2016. "Individualism, Neoliberalism and Heroism: An Auto-Ethnographic Account of Pro-Environmental Transitions." Honours Thesis, Trent University, Peterborough, ON.

Graham, Peter, Cassandra Kuyvenhoven, Rena Upitis, Adeela Arshad-Ayaz, Eli Scheinman, Colin Khan, Allison Goebel, R. Stephen Brown, and Alice Hovorka. 2020. "The Emotional Experience of Sustainability Courses: Learned Eco-Anxiety, Potential Ontological Adjustment." *Journal of Education for Sustainable Development* 14, no. 2 (September): 190–204. https://doi.org/10.1177/0973408220981163.

Irfan, Umair. 2019. "Greta Thunberg Is Leading Kids and Adults from 150 Countries in a Massive Friday Climate Strike." *Vox*, September 20, 2019. https://www.vox.com/2019/9/17/20864740/greta-thunberg-youth-climate-strike-fridays-future.

Kuyvenhoven, Cassandra, and Peter Graham. 2017. "Exploring the Experiential Dimension of Sustainability Courses, School of Environmental Studies, Queen's University." Presentation at The Showcase on Teaching and Learning, Centre for Teaching and Learning, Queen's University, Kingston, ON, May 3, 2017.

McKinley, Andrew. 2008. "Hope in a Hopeless Age: Environmentalism's Crisis." *Environmentalist* 28 (April): 319–26.

Wiek, Arnim, Angela Xiong, Katja Brundiers, and Sander van der Leeuw. 2014. "Integrating Problem- and Project-Based Learning into Sustainability Programs: A Case Study on the School of Sustainability at Arizona State University." *International Journal of Sustainability in Higher Education* 15 (4): 431–49. https://doi.org/10.1108/IJSHE-02-2013-0013.

Willison, John, and Kerry O'Regan. 2007. "Commonly Known, Commonly Not Known, Totally Unknown: A Framework for Students Becoming Researchers." *Higher Education Research & Development* 26 (4): 393–409.

13

Sustainability Pedagogy
Keeping Up with Millennials and Generation Z

KOUROSH HOUSHMAND

FROM MY VANTAGE POINT as a millennial who has recently graduated from university, sustainability education is often misrepresented by antiquated pedagogy. In my educational experiences, curriculum has often overemphasized sustainability as environmentalism, and I believe this in turn has undermined the full scope of the sustainability sciences. This is not to deny that environmentalism is important for sustainability; it is to say that sustainability is much more than environmentalism. When I speak about sustainability, I adopt the "three E's" model—environment, economics, and equity—where an interdependence between these three realms can support sustainable development (Caradonna 2014). I argue that sustainability as a theoretical framework is facing an incredible branding and economic transformation. It has become the core tenet of a branding renaissance for consumer products geared toward a millennial audience. According to the 2015 Nielsen report, 66 percent of global respondents are willing to pay more for sustainable goods, but this is even higher for millennials, at 72 percent. The new millennial consumer reveals the growing disconnect between sustainability practices and sustainability education in post-secondary institutions. In this chapter, I share how the overall idea of sustainability science education can and should be made more relevant for my generation and for

Generation Z, which includes those who were born in 1997 or later, according to the Pew Research Center (Dimock 2019).

My opinions in this chapter are largely informed by my experiences as a student who was fortunate to observe and advocate for many policy decisions in Ontario education. I started working in sustainability and education pedagogy from the age of fifteen. My relationship with education started when I was a student trustee at the Toronto District School Board, where I had the opportunity to represent over 260,000 students. As a student representative, I've had opportunities to engage with premiers, education ministers, school boards, energy sector executives, and international experts in education. Ever since, I have been published nationally and internationally for my advocacy on K–12 and post-secondary education issues. I consider myself a lucky observer of sustainability and education policy, not a full-fledged expert. Importantly, I do not believe there is a single "student voice," and I am not under the illusion that I speak on behalf of an entire demographic; rather, I am just one voice among many.

My opinions have also been informed by my experience as a young entrepreneur in the solar industry. When I was seventeen, I started a non-profit organization called Solar for Life, which worked on solar solutions for off-grid areas in the most peripheral parts of the world. Our material impact was core to what we did, but more importantly, my goal has always been to make uninteresting things more interesting for certain audiences. Solar for Life was an attempt to make solar energy cool and relevant for a generation that I believe truly cares about sustainability.

While many of my past experiences inform the things I know, they also shed light on the things I do not know. I was raised thinking about education pedagogy and sustainability at a time when free online instruction began competing with university curricula around the world. As a result, my view of traditional education pedagogy is both critical *and*, admittedly, limited to my experiences growing up in the "open-sourced" world of information. My goal with this chapter is not to paint an overly somber picture. Surely, sustainability education confronts a huge burden to maintain relevance as an identifiable field of study. It also faces an equally large, if not larger, opportunity to expand its academic appeal.

My assertion is rather simple: sustainability, instead of being approached as an independent field of study, should serve as a framework for all applicable disciplines. In other words, sustainability education ought to be treated like basic literacy: a prerequisite for everything else one learns. Isolating sustainability from the various issues that it affects undermines what I believe is the essence of sustainability: an overarching framework for the social and economic choices that we make.

Sustainability education has been transformative in the sense that millennials—the new consumers I referenced—have grown up in some places, like Ontario, to be the first generation in Canada to have learned about sustainability in K–12 schooling through a broader, non-environmental context. For example, sustainability was part of Ontario's curriculum for grade 7 and 8 history and geography in the early 2000s (Ontario Ministry of Education 2004). These millennials are now post-secondary students, and more importantly, they are new economic purchasers. Sustainability is currently an important component in Ontario's curricula for grade 7 and 8 geography and grade 11 and 12 science (Ontario Ministry of Education 2018, 2008), among its other curricula, and thus continues to influence education with the same impact for Generation Z.

Herein lies the opportunity: the new generation of thinkers and buyers now looks at sustainability through a broader economic lens, and thus pedagogy must reflect this new reality. This economic lens also encompasses a distinct understanding of sustainability as a nuanced tool for innovation, branding, and pricing. Out of my subjective experience as a millennial who works in a current job where I examine how people consume information and make decisions, I have chosen three cases that I believe exemplify the new reality of sustainability for young adults and middle-aged consumers. I argue that these market realities must increasingly guide curricula in post-secondary institutions. Of course, my analysis is not meant to pinhole sustainability education in these business-forward frameworks, but to share a newly relevant context for its pedagogy. Simply put, the millennial generation constitutes a class of economic purchasers that makes consumer decisions based on different considerations—sustainability-oriented ones (Dabija, Bejan, and Vasile 2019; Naderi and Van Steenburg 2018). As a result, for sustainability pedagogy to remain relevant to this new generation, I believe it needs to fully embrace what sustainability means to the lives of millennials, who are in their biggest purchasing-power stage of life (Hoffower and Kiersz 2021). The three types of companies profiled below serve as examples for linking sustainability to important consumer trends in the millennial and Generation Z populations. Consumer behaviour is one of our most elemental connections to sustainability systems and broadly links environment, economics, and equity.

Airbnb as Sustainability Technology

Some of the most incredible technological innovations have been made with sustainability as a core tenet of their innovative processes. Airbnb, for example, is not simply a marketplace for travel accommodations, but has completely transformed what sustainable spaces look like—few technologies have transformed the concept of shared living spaces as much as Airbnb. It has democratized the hospitality industry by instilling a system of trust among strangers at a significant scale. Part of Airbnb's branding strategy is also a deep-rooted understanding of the new appeal to their audience, largely made up of young urban professionals and millennials (Lock 2019). Their mission is to reimagine the way people live and belong:

> Decades ago, travellers stayed in boarding homes, neighbours shared what they had, and ordinary people powered the economy. These activities are re-emerging through a new movement called the sharing economy, where everyone can participate…We imagine a more accessible New York that even more people can afford to visit, where extra space in people's homes will not go to waste, and where millions of visitors patronize neighborhood small businesses across all five boroughs. (Airbnb 2013)

Could one make a case that Airbnb is a sustainability-minded company? This is debatable—while Airbnb is part of the sharing economy, others argue its sustainability is questionable, given the job losses for those in the hotel industry. But should Airbnb's impact on sustainability be taught in post-secondary institutions? In my view, yes. In fact, Teubner

(2014) argues that it is one of the ten ideas that will change the world in the twenty-first century. This is the junction point where sustainability education becomes relevant to technological innovation and industries alike. Airbnb (2018) affects sustainability, as its home sharing listings typically generate "less waste, use less water, and drain less energy than traditional hotel accommodations." This is not meant to validate Airbnb's commitment to sustainability, but rather to provide an example of how sustainable practices can occur in multifaceted ways. It is important for schools to see this as a case study for sustainability in businesses. In this way, sustainability ought to be taught as an underlying condition for successful innovations and business practices, and it ceases to be just another interdisciplinary field.

The example of Airbnb is simply meant to reinforce the notion that we must teach sustainability with a broader scope in curricula. While being its core tenet, sustainability is not just about environmentalism; it can also be about a company's bottom line and business model.

Cold Pressed Juice as Sustainability Pricing

Cold pressed juice is another product with a strategy deeply rooted in this wave of sustainability branding, and sellers are able to command a US$9 price point for a 300-ml bottle of natural juice. I focus on cold pressed juice because it is part of a larger trend of millennials paying premium prices for food products they believe are produced in a more ecologically and socially responsible way (Angus 2018). Cold pressing is a juice-making process that extracts more out of the fruit and requires fewer energy inputs during production than conventional processes, but it is more expensive. This is a sustainability-angled product, and while its impact is not vast and not everyone purchases cold pressed juice for sustainability reasons, it is an example that brings to light a new generational pulse for sustainability pricing when it comes to products. It does not take an economist to see that pricing equilibrium now incorporates a general willingness, from a certain demographic of consumers, to pay higher prices for more sustainably sourced goods. And while there may be other factors, such as health benefits, that justify the price of cold pressed juices and similar products for consumers, sustainable branding is also an important element that nudges a pricing expectation. There is also an opportunity here to invite students to discuss the consistencies between pricing expectations and ethical pricing.

Farm-to-Table as Sustainability Branding

An even more popular example of sustainability branding is the culinary wave of farm-to-table restaurants that have reimagined what the consumer-facing side of food sustainability can look like. Farm-to-table foods first gained popular footing in San Francisco, a city that often foreshadows future innovation and global urban trends, but they have also become more prevalent across other North American cities, including Toronto and Vancouver. Farm-to-table restaurants, which shorten the link between the farmer and the consumer, feed into a growing public awareness of the environmental and social ramifications of food production, distribution, and consumption.

Millennial consumers en masse see sustainability branding on products as a selling point, enough so that they are willing to pay more for it. For example, a study by the Hartman Group, called "Understanding Millennials," indicates that millennials are willing to pay more for farm-to-table experiences (Taher 2014). Millennials are also willing to pay more for food believed to be ethically sourced and produced, and this commitment continues in Generation Z (Nielsen 2015). I do not use the word "branding" in a negative light. I simply use the term out of convenience to describe how consumers like the idea of sustainably produced and labeled items and services. Whether or not sustainability branding is seen as a fad, educators can observe the trends and encourage students to critically engage with the drivers and impacts of sustainability products.

The Way Forward

Sustainability has somewhat successfully been integrated into market dimensions for a millennial audience and positioned as a variable for pricing, branding, and (in the case of Airbnb) technology innovation. Business programs should integrate sustainability concepts into their existing curricula (Barber et al. 2014). Topics such as consumer ethics can inform students about what guides many consumer behaviours and why consumer values related to sustainability have disrupted traditional pricing equilibriums (Carroll and Buchholtz 2014). I believe that students could learn about sustainability more successfully if the pedagogical framework in academic institutions demonstrated the relevance of sustainability to a millennial-driven economy. Students would benefit from understanding sustainability in the context of phenomena like Airbnb's technology, the pricing of cold pressed juices, and farm-to-table branding.

Out of my humble experience, I have also suggested that marketing should teach sustainability branding. A sustainability lens should be brought to curricula on technological innovation, as I have suggested for the rise of Airbnb. Pricing classes should include lessons on consumer values tied to sustainability, as demonstrated by the increased purchasing of cold pressed juices. My point is that in order to bridge a gap and respond to a new window of opportunity for learning that is relevant for students, sustainability education must adapt to the speed and relevance of sustainability in terms of what is relevant to the lives of millennials and those in Generation Z. Post-secondary institutions have the power—if not the obligation—to utilize the demand for sustainability and turn what are sometimes fads in the branding world into content for deeper learning. More than ever, young people are engaged in sustainability—I believe it is the job of the education system to capitalize on this opportunity in ways that resonate with students.

There is an increased market demand to teach sustainability through a broader lens of socio-economic consumer choices and opportunities for innovation—not just through an environmental narrative, but also in a way that sustains income, human dignity, inclusivity, creativity, and resilience. In other words, young people care about sustainability broadly as a concept, and academic approaches need to tap into the heart of what sustainability means for them. I acknowledge that it is difficult to devise how this new application of sustainability could be taught in higher education. An adaptive approach would require a relatively significant transformation from the way sustainability has been approached pedagogically thus far to achieve a more nimble and responsive pedagogy.

Bearing in mind these challenges, post-secondary institutions should strive to offer sustainability education through an interdisciplinary lens (Bernstein 2015). This also means integrating sustainability concepts more deeply across the curricula of different faculties. It should become a core competency in every academic discipline, not be an independent field of study. In this book, Boone's chapter in particular speaks to various core competencies, such as systems and collaborative thinking, that each discipline should take on as they revise and develop courses. I see a future where sustainability education is the bedrock of learning about various fields and threads as tendrils across fields. In my view, sustainability teaching has too often been a continuation of a post-secondary and K–12 education model that is not responsive to the current context. This is not a novel argument but rather a simple reiteration of—or perhaps a cry for—a change that is much needed to maintain the pedagogical relevancy of sustainability education.

This chapter has attempted to invite those in higher education to view sustainability as a lens through which other fields ought to be taught, rather than a field of study. A new generation of academics and consumers ought to be taught in radically different ways to realize the interdisciplinary applications of sustainability. I have offered a few topics here that could enrich sustainability education for millennials and those in Generation Z. Young people care about sustainability, and so sustainability pedagogy needs to keep up with its real-world applications for a millennial and Generation Z audience.

References

Airbnb. 2013. "Who We Are, What We Stand For." *Airbnb* (blog), October 3, 2013. https://blog.atairbnb.com/who-we-are/.

Airbnb. 2018. "How the Airbnb Community Supports Environmentally-Friendly Travel Worldwide." *Airbnb Newsroom*, April 19, 2018. https://news.airbnb.com/how-the-airbnb-community-supports-environmentally-friendly-travel-worldwide/.

Angus, Alison. 2018. *The Top 10 Global Consumer Trends for 2018*. London: Euromonitor International.

Barber, Nelson A., Fiona Wilson, Venky Venkatachalam, Sara M. Cleaves, and Josina Garnham. 2014. "Integrating Sustainability into Business Curricula: University of New Hampshire Case Study." *International Journal of Sustainability in Higher Education* 15 (4): 473–93. https://doi.org/10.1108/IJSHE-06-2013-0068.

Bernstein, Jay H. 2015. "Transdisciplinarity: A Review of Its Origins, Development, and Current Issues." *Journal of Research Practice* 11, no. 1 (article R1). http://jrp.icaap.org/index.php/jrp/article/view/510/412.

Caradonna, Jeremy L. 2014. *Sustainability: A History*. New York: Oxford University Press.

Carroll, Archie B, and Ann K. Buchholtz. 2014. *Business and Society: Ethics, Sustainability, and Stakeholder Management*. 9th ed. Boston: Cengage Learning.

Dabija, Dan-Cristian, Brândșua Mariana Bejan, Dinu Vasile. 2019. "How Sustainability Oriented is Generation Z in Retail? A Literature Review." *Transformations in Business & Economics* 18 (2): 140–55.

Dimock, Michael. 2019. "Defining Generations: Where Millennials End and Generation Z Begins." *Pew Research Center*, January 17, 2019. https://www.pewresearch.org/fact-tank/2019/01/17/where-millennials-end-and-generation-z-begins/.

Hoffower, Hillary, and Andy Kiersz. "The 40-Year-Old Millennial and the 24-Year-Old Gen Zer Are in Charge of America Right Now." *Business Insider*, September 26, 2021. https://www.businessinsider.com/24-gen-z-trends-40-millennial-spending-changing-economy-2021-9.

Lock, S. 2019. "Airbnb Users by Age Group in the U.S. and Europe 2017." *Statistica*, August 9, 2019. https://www.statista.com/statistics/796646/airbnb-users-by-age-us-europe/.

Naderi, Iman, and Eric Van Steenburg. 2018. "Me First, Then the Environment: Young Millennials as Green Consumers." *Young Consumers* 19 (3): 280–95. https://doi.org/10.1108/YC-08-2017-00722.

Nielsen. 2015. *The Sustainability Imperative: New Insights on Consumer Expectations*. October 2015. https://www.nielsen.com/wp-content/uploads/sites/3/2019/04/global-sustainability-report-oct-2015.pdf.

Ontario Ministry of Education. 2004. *The Ontario Curriculum: Social Studies, Grades 1 to 6; History and Geography, Grades 7 and 8, 2004*. https://www.uwindsor.ca/education/sites/uwindsor.ca.education/files/curriculum_-_social_studies_1-6_history_geography_7-8.pdf.

Ontario Ministry of Education. 2008. *The Ontario Curriculum, Grades 11 and 12: Science, 2008*. http://www.edu.gov.on.ca/eng/curriculum/secondary/2009science11_12.pdf.

Ontario Ministry of Education. 2018. *The Ontario Curriculum: Social Studies, Grades 1 to 6; History and Geography, Grades 7 and 8, 2018*. http://www.edu.gov.on.ca/eng/curriculum/elementary/social-studies-history-geography-2018.pdf.

Taher. 2014. "What Millennials Crave at Restaurants." *Taher Inc.* (blog), September 9, 2014. http://www.taher.com/millennials-crave-restaurants/.

Teubner, Timm. 2014. "Thoughts on the Sharing Economy." In *Proceedings of the International Conferences ICT, Society and Human Beings 2014, Web Based Communities and Social Media 2014, e-Commerce 2014, Information Systems Post-Implementation and Change Management 2014, and e-Health 2014*, Vol. 11, edited by Piet Kommers, Pedro Isaías, Claire Gauzente, Miguel Baptista Nunes, Guo Chao Peng, and Mário Macedo, 322–26. Lisbon, Portugal: IADIS.

Conclusion

NAOMI KROGMAN

SUSTAINABILITY EDUCATION, tied to either surviving or thriving, will continue to be a deep commitment for universities. The authors of this volume summarized numerous reasons for this, but key among them is that millennial and Generation Z students recognize that they will need to collaborate to build and re-build the very systems we rely on for basic needs, human dignity, and sustainable livelihoods. Climate change is forcing societies to confront the risks for nature and humanity. Increased awareness of inequalities and discrimination in the US and Canada has forced institutions to address systemic outcomes in relation to power, justice, and well-being, which are woven into collective decisions about land and sea. These are easily seen regarding Indigenous People's rights and governance to protect their traditional territories. Issues of power, justice, and well-being are tied to the quality of and access to food, as is readily observed in the food deserts of North American cities and in the cost of a nutritious diet for the working poor. These issues are also bound to the suffering of vulnerable populations, as observed from looking at who is most likely to die during a pandemic, natural disaster, or toxic industrial accident.

Students want to be part of the solutions and figure out how to prevent injustices in the first place, rather than just cleaning up the mess or offering minimum amelioration. I see students willing to make sacrifices in their own lives to avoid banking their happiness on materialism and status, and to confront their own confirmation biases and invest in the communities to which they belong. In my experiences as a professor for more than twenty years and as a dean of a Faculty of Environment for three years, I have seen students who are increasingly willing to roll up their sleeves for a long-term and tenacious commitment to change the status quo. This makes sustainability education all the more appealing to this younger population, but also to non-traditional students who are seeking to be part of the changes they wish to see, especially in light of persistent injustices such as racism and the perpetuation of economic systems that make life precarious for so many. While not the case in Canada (Grenier 2020), there are record numbers of millennials running for office in the United States (Serai 2020). I hope that advances in sustainability education will propel more of our young adults to lead sustainability changes through political leadership.

Students are increasingly calling for the authentic recognition of Indigenous land rights (e.g., through the Idle No More movement and other decolonial movements in post-secondary institutions) and for settler–ally commitments to address the wrongs and continuing unfair practices in police, social welfare, education, and resource management systems. They call for respect for Indigenous traditional knowledge and Indigenous interests, and for talent and intelligence in politics, music, film, the criminal justice system, and educational content. Sustainability education calls for more Indigenous content—built into the curriculum rather than offered as stand-alone courses—and for, as Megginson (chapter 5) and Perkins (chapter 11) suggest, more partnerships between tribal colleges, Indigenous communities, and other higher education institutions.

Most of the jobs in the near future will require workers to integrate sustainability thinking into the decisions they make, in part to save money, but also to build efficiencies into infrastructure (e.g., green infrastructure to minimize flooding in cities) and policy (e.g., regulations such as phased-in requirements for the electrification of vehicles), so that climate change mitigation and adaptation are accomplished at the same time. Universities have a key role to play in preparing future workers to respond to resource scarcity, persistent poverty, technological accidents, civil unrest, and disasters (especially floods, droughts, forest fires, and tropical storms, which are occurring with increasing frequency).

The contributors of this book speak to the challenges noted above, providing unique perspectives. These distinctive voices—which include professors, academic administrators, subject area experts, practitioners, and two recent graduate students—provide valuable insights on the future directions and opportunities of sustainability education at North American universities, as well as the strategies and tactics that can help us progress toward those futures. Despite the different social locations of the contributors, some common themes emerged. These themes are summarized in the recommendations below.

Invite Students to Grapple with Grand Challenges

To take on grand challenges, such as climate change and persistent poverty, several contributors advocate teaching critical thinking to get to the core of the problem. This calls for exercises that train students to apply reason and use reflection to ask the right questions and then use the best methods, analyses, and contextual application of knowledge to get the best and most informed answers. Dietz (chapter 7) holds that we not only need to get the science right, but we also need to get the right science(s) involved to address grand challenge questions. Malcom (chapter 9) suggests teaching students to start with the questions, "What is your problem?" and "What do you want to understand?" Dietz (chapter 7) also contends that we should familiarize students with varying levels of uncertainty so they truly embrace the adaptive, trial and error nature that will be required to make improvements over time (and to avoid silver bullet, singular solution thinking). Dale (chapter 8) emphasizes that there are no single right answers to wicked problems, but there are better informed answers that can lead to improvements.

Several authors prompt students to ask the big questions by challenging them to identify their own values and those of society and then determine where policies, institutional practices, and social and ecological conditions appear to be antithetical to those values. The natural question that follows is why do we accept that certain conditions persist, such as the continuation of coal mining? Dietz (chapter 7) argues that we need to understand the role that values (not just facts) play in decision making, particularly the values of those who make decisions and those who are impacted by them. Steelman (chapter 10), Dale (chapter

8), and Perkins (chapter 11), in particular, call for placing students in the middle of wicked problems, where they can circle the problem from various stakeholder positions, understand the values behind different viewpoints, and seek locally sensitive, equity-enhancing solutions.

Try Out Interventions
Similarly, several authors offer guidance on how to use science, evidence, and community capacity to not only understand a problem but also discern how to intervene. Epp (chapter 2), Dietz (chapter 7), and Boone (chapter 3) call for skills development to form the basis of an experimental approach to assess the uncertainty of intervention outcomes, avoid path dependencies, and design flexible interventions.

Dale (chapter 8) invites students to do "challenge-based" projects that are tied to wicked problems. In Royal Roads University's certificate program in sustainable community development, for example, current sustainability challenges are identified—in partnership with local governments, civil society leaders, and the private sector—to provide students with real-world experiences. Dietz (chapter 7) also holds that it is important to teach students the connections between fundamental knowledge and applied research, where experiments foster social learning for sustainability. Goebel (chapter 12) describes her experiences using inquiry-based learning—a self-directed, step-by-step process where students develop questions related to their topics, try to "do something about it," and then reflect on the emotions, incentives, and barriers they faced, as well as the difference they would like to make. Goebel writes that, as a result of this process, some students identify the limits of individual action and realize the need for collective action.

Incorporate Innovation
Dale (chapter 8), Houshmand (chapter 13), Sharpe (chapter 6), and Dietz (chapter 7) call for university teachers to be innovative in the topics and case studies they teach and to keep pace with the rapid technological changes that will influence human life, such as the rise of big data, artificial intelligence, biotechnology, and the role of social media in learning and communication. Houshmand (chapter 13) suggests that academe could go much further to foster entrepreneurial curiosity and market acumen around products viewed as more sustainable by

millennials and those of Generation Z. Dale (chapter 8) calls for professors to apply social learning (where participants invite trial and error experiences and learn from each other within communities of practice) and co-teach courses with other disciplinary and practitioner experts to continually upgrade sustainability content to respond to change and current grand challenges. The notion of "failing forward" occurs to me here, where professors and students are asked to move out of their comfort zones to examine sustainability experiments with policy and practice that are in contrast to current social practices. Popular social experiments might include the sharing economy, co-housing, and virtual communities that support ethical and sustainable practices.

Sharpe (chapter 6) calls for significant shifts in academia to expand teaching on how technology moves through the stages of the innovation chain. She contends that sustainability education should go beyond its typical focus on research and development to encompass the whole innovation chain, including research, development, demonstration, deployment, and diffusion into the market (RD4). Instructors need to teach students how to leverage growth capital for all five stages of RD4 to understand how clean technology, for example, moves through the stages of innovation to commercialization. Sharpe argues that academia needs to be better at providing testing grounds, such as having appropriately designed laboratories that can mimic field conditions and be customized for different applications.

Boone (chapter 3) contends that innovative thinking is fostered by training in the arts, humanities, and design, as these disciplines inspire students and prompt reflection. He reminds us that the arts help us understand the human condition and motivators for human behaviour. For example, Joan Greer, a professor in the Department of Art and Design at the University of Alberta, teaches a novel course on the visual culture of natural history (with an emphasis on botany and entomology) and early environmentalism, which brings together knowledge of history, nature, and art. Artwork by Greer's students was displayed in the room at our workshop in Banff, where all authors of this book initially shared ideas for their chapters. This led to interesting conversations about the representation of landscapes, which reminded us of the power of art to transport us to a different time and place and emotionally engage us with the future of landscapes.

Perkins (chapter 11) calls for higher education to provide more in the curriculum on people's innovations to share risks and protect each other through cooperatives, mutual aid, utopian communities, and grassroots collaborative economic initiatives. Goebel (chapter 12) affirms that focusing on success stories and collective action experiments offsets the doomsday thinking that many students have been stuck in, due to the bad news they've heard about the environment their whole lives. There is an opportunity here to teach hopeful innovations, for example, that address some key thematic areas from COP26 (held in the UK in 2021)[1] on clean energy, zero emissions vehicles, green financing, nature-based solutions (e.g., cleaning wastewater by protecting wetlands), adaptation, and resilience. For example, educators could teach about social innovations of collaboration and cooperation to organize more effectively to address climate change through activities such as community gardens and co-housing.

Develop Holistic, Systems-Based Curricula

To take on grand challenges, investigate interventions, and analyze innovations, several authors call for more holistic, place-based, and systems-based curricula. Boone (chapter 3) refers to this as an integrated systems approach, where issues like local wetlands are viewed across time and space, and within different decision-making spheres. Perkins (chapter 11) also argues that watersheds are ideal subjects to teach systems thinking and commons governance. Dietz (chapter 7) argues that one way to teach systems thinking is to teach different approaches to decision making for sustainability, as this inherently requires applying different disciplinary approaches to incorporate values and an understanding of the various forces (psychological, social, political, and economic) that lead to uncertainty in outcomes.

Dale (chapter 8) also emphasizes uncertainty and the consequential need for holistic and systemic thinking. To do this, she suggests that departments could be organized around challenge areas rather than disciplines. Similarly, Malcom (chapter 9) suggests that universities should offer problem-based learning that fosters depth and breadth in student learning and concomitantly rewards professors for "border crossing." Boone (chapter 3) offers the notion that border crossing can sometimes be fostered by temporarily relocating a faculty member to another department to build courses and research programs together more fluidly. Finally, Dale (chapter 8) suggests co-teaching and co-learning approaches to make continual updates to sustainability curricula.

Common property resource management (Boone, chapter 3 and Perkins, chapter 11) and polycentricity (Perkins, chapter 11) are excellent topics to teach holistic and systems thinking, as well as the levels of uncertainty inherent in dynamic social and ecological systems. An advantage of commons governance scholarship is that it includes equity issues and often illustrates the power of participatory democracy and local responsibility in the sustainable management of resources. Similarly, Dietz (chapter 7) implies that by asking students to study a problem that moves across scales (e.g., individual to household to local to regional to national to international), they can learn to apply different types of knowledge to the same problem, with particular sensitivity to the complex challenge of a successful management approach in a local context.

Megginson (chapter 5) argues that mathematics, the "language of science," is essential in sustainability training, especially to understand rates of change and change interactions, and he outlines the math training that should be part of two- and four-year sustainability-related degrees and graduate programs. More specifically, Megginson explains that many factors interact in the behaviour of natural systems, and a standard course in partial differential equations enables students to have critically important math skills to understand transitions in ecosystems. In his survey of sustainability-related programs in US and Canadian tribal colleges, however, he notes that many of these programs lack sufficient mathematics requirements, especially when it comes to calculus.

Several authors also call for greater efforts to develop clear pathways of learning in sustainability education, where learning outcomes are identified and staged throughout academic programs. Malcom (chapter 9) speaks to the problem of isolated courses, or coursework not built together to tell a story, and inadequate guidance to help students navigate across course offerings. Boone (chapter 3) offers a clear set of key competencies for sustainability in higher education (based on Wiek, Withycombe, and Redman 2011), which Arizona State University used to develop program-level learning outcomes. Steelman (chapter 10) suggests that sustainability programs should be built on the basis of clearly identified learning outcomes.

Learn Transdisciplinarity through Community-Engaged Research

Multiple authors recommend teaching transdisciplinarity. I especially appreciate Dale's (chapter 8) definition of transdisciplinarity, which comes from Max-Neef (2005, 15):

> Discipline and trans-discipline must be understood as complementary. The transit from one to the other, attaining glimpses from different levels of reality, generates reciprocal enrichment that may facilitate the understanding of complexity.
>
> Transdisciplinarity, more than a new discipline or super-discipline is, actually, a different manner of seeing the world, more systemic and more holistic.

To really understand different levels of reality, many scholars in this book recommend experiential learning in general, and more specifically, community-engaged research or service learning. This recommendation is different from the call for more teaching on systems and holistic thinking. The authors distinctly call for students to be engaged in experiences and research outside of the classroom, with societal knowledge holders and practitioners who can share with them on-the-ground strategies for building solutions, troubleshooting, communicating with stakeholders, and operating in a specific cultural and political context. Steelman (chapter 10), for example, points out how students learn how to collaborate and co-produce knowledge by being involved in community-engaged research. Similarly, Perkins (chapter 11) and Goebel (chapter 12) argue that team building skills are essential. Dale (chapter 8) holds that transdisciplinarity must not only be taught but also modelled, showing how people transcend disciplines, political views, cultural differences, divergences in skills, and ties to place to collaboratively achieve improvements over the status quo. Sharpe (chapter 6) implies that students need to learn a pragmatic mindset, where taking action empowers people and there is confidence in finding a coalition of the willing.

Several authors call for deeper work with those who live close to the land and problem solve on the ground. For example, Epp (chapter 2) draws our attention to the concept of "phronesis," which refers to "practical wisdom shaped by experience and marked by a capacity for judgment." He stresses how important it is for researchers and students to work side-by-side with those who have long-standing experience with learning a social–ecological system. Boone (chapter 3) maintains that sustainability programs need to be relevant to external communities, and this requires regular communication between university partners and instructors. Faculty need to get comfortable, Malcom (chapter 9) writes, with presenting to non-specialist audiences, and universities need to reward faculty for "border crossing."

Dietz (chapter 7) holds that we need to teach students the important nexus between analysis and deliberation and to confirm with decision makers that the science offered to them answers their questions. Steelman (chapter 10) advocates for a more concerted effort among professors to link their research to public deliberation on sustainability issues. Several authors share their enthusiasm for the suite of options to foster these kinds of experiences. Options such as internships, service learning, off-campus undergraduate research, and project-based courses that might use flipped classrooms (where students do a project with outside communities and then come together with their classmates in a physical classroom to discuss what they are learning over the term) blend application of knowledge and skills with hands-on experience.

Create Places for Debate and Deliberation

For all the contributors to this book, the ivory tower of academe, where professors provide the evidence and leave others to figure out how to apply it, is a chapter of university history to leave behind. Given the special place the academe has in society, for example, Steelman (chapter 10) suggests that universities are ideal for creating safe places for democratic debate and especially for bringing those who feel disenfranchised into deliberative dialogue. Steelman and Epp (chapter 2), in particular, hold that universities need to create spaces to address controversial topics. For example, these topics could include how to create resource management institutions that respect Indigenous rights and ways of knowing, how to envision urban development and re-development that seek to eliminate racial, gender, sexual, and religious discrimination, and how to plan for sustainable livelihoods that allow all people a living wage.

Many of the authors argue that a key part of sustainability education is learning how democracy could and should work, learning how to do critical stakeholder analysis of what constitutes the "public good," and developing citizen engagement and leadership skills. Students need to understand how power operates to inform how they navigate their own sustainability efforts, confront power over time and from different angles, and build allies across interests and positions. Sustainability education should not shy away from the political nature of social change.

Be Responsive to Student Interests

Student interests in sustainability shift over time, and universities need to adapt to building these new interests into their academic programming and public engagement. Houshmand (chapter 13) and Boone (chapter 3) call for universities to be nimble, incorporate learning experiences in emerging areas of interest, and support students who organize around issues of concern. Universities can seize greater opportunities to play to student strengths in using social media, networking, and employing other online communication to build community and pursue social entrepreneurship. Steelman, in particular, calls for instructors to continually update their learning modalities to those most appealing and effective for students. We expect far more adoption of new learning modalities to come out of the widespread experience of remote teaching during the pandemic.

Perkins especially argues that young people are interested in connections between pedagogy, activism, and research. Local controversial issues can be taken up in place-based education, which, as Epp argues, students find particularly relevant. Finally, Jones (chapter 4), Bergstrom (chapter 1), and Boone (chapter 3) indicate that universities can pique students' interest in sustainability by developing campus as a living lab initiatives, sustainability internships, new introductory courses, embedded and professional certificates, and minors in sustainability. Bergstrom notes that universities are currently making progress in many of these areas, but that these initiatives should be more holistically integrated throughout the systems of higher education institutions. While not addressed in this book, there is significant opportunity to develop more post-baccalaureate certificates and offer other micro-credentials to foster lifelong learning in sustainability in the broader population.

Change Reward Structures at Universities to Incentivize Sustainability Education

Incentive structures at many universities do not support inter- or trans-disciplinary teaching and research, and these structures need to change. Dietz (chapter 8), Boone (chapter 3), and Steelman (chapter 10), for example, contend that faculty should be rewarded for their research and teaching talents to search for solutions that are practical and understandable, and that will have near-term positive effects. Malcom (chapter 9) argues that universities need to create rewards for "border crossing," i.e., working with non-specialist audiences. Several authors hold that universities will ultimately need to modify tenure and promotion guidelines to reward sustainability scholarship and teaching and stop the "publication for the smallest sliceable unit" that is incentivized in many "publish or perish" systems. Other forms of scholarship, such as reports, documentaries, policy briefs, manuals, and other digital media, need to be recognized as important research products that have far more reach and usability than refereed journal articles. We need to find better ways in higher education institutions to track impacts of scholarship, given the continued lower value that research-intensive universities place on these other forms of scholarship.

It is important to acknowledge that some faculty are likely to resist learning about and teaching sustainability, and to resist changes to their reward systems, given the traditions and cultures of their disciplines that have laid out what generally constitutes core content for a particular course, as well as academic credibility and excellence (Peet et al. 2004). For some disciplines, for example, especially those in science and engineering, value-laden choices that guide what science is done, for what purpose, and as applied to which sustainability problems are not part of the course. Additionally, professors may not have the expertise and experience required to implement sustainability-related teaching and learning (Jones, Selby, and Sterling 2010; Nicolaides 2006; Thomas 2004), or they may resist it because they see it as part of a "moralizing agenda" that they don't support (Butcher 2007). In some cases, professors may see sustainability as having no relevance to their courses or their discipline (Dawe, Jucker, and Martin 2005), or they may not be interested in expending the effort it takes to incorporate sustainability into their teaching and research (Lozano 2010). In sum, it is important to recognize that it will require internal deliberation for faculty leadership and advocacy to change the incentive structures from traditional outputs (refereed journal articles and books) to outputs that are tied to changes in thought, practice, and policy in sustainability teaching and scholarship. University administrators, from contributors to this book, are called to foster internal discussions about the overall evolving purpose of disciplines and post-secondary education in society.

In my own Faculty of Environment at Simon Fraser University, many of these barriers have been reduced in the School of Resource and Environmental Management, where faculty are supported in their efforts to integrate ecology, economics, and social science into most of the course offerings. They are also rewarded in their biennial reviews for conducting problem-based collaborative research and producing user-friendly research products, alongside scholarly works, for governments, First Nations, foundations, industries, and non-profit entities.

Conclusion

Many incremental changes are happening in universities to advance sustainability education and research in North America. However, we could still ask, "So what?" After all, universities are always changing to some degree and incorporating new theories, science, scholarship, and practical knowledge into their curricula.

Sustainability education is needed to develop knowledge, skills, and, importantly, the imagination of a better future—one where the human species not only survives but thrives as part of an overall healthy living system. Universities need to hold up a mirror to the activists, community organizers, volunteers, writers, artists, actors, reporters, musicians, and poets that provide an interstitial and personal understanding of the values and embodied experiences at stake as our living systems change or decline. Universities need professors and instructors to show an affability toward applying appropriate disciplines to wicked problems, listening to and respecting divergent positions on how to define problems and advance solutions, and building collaborative approaches for short- and long-term solutions. Sustainability educators can better liaise with the private sector to pilot and scale up solutions. While advancing technological and other innovations, they can continue to teach students the role of regulations and successful common property resource management practices (which are often overlooked due to the hype around technological solutions).

Dale (chapter 8) points out that there are projected to be over 260 million students worldwide by 2025 (Maslen 2012). We must recognize, as Bergstrom (chapter 1) indicates, that universities educate most of the people who develop and manage society's institutions (ULSF 2015). Most decision makers in government, industry, and civil society have earned a bachelor's degree. Given the increasing numbers of women and people of colour in academe, we can also predict, as Jones (chapter 4) points out, that more students will be drawn to sustainability, given their greater likelihood of being interested in this subject as it relates to equity-enhancing change, and because more relatable role models will increasingly stand before them in the classroom.

Many contributors to this book told me they know sustainability education makes a difference because they have been doing it for two or more decades as educators, thought leaders, and administrators. They have witnessed previous students going on to teach advances in sustainability to other students, as professors, instructors, trainers, consultants, industry leaders, government leaders, activists, and professionals. They have also seen students in varied roles, as knowledge brokers, mobilizers, and translators. Universities are one of the few learning places, as Bergstrom (chapter 1) points out, where research and teaching are deliberately designed to promote the public good. Universities are bedrocks for teaching about sustainability, given the continued call for research and partnerships that promote equity, inclusivity, justice, democratic processes, and social and ecological health. As Epp (chapter 2) suggests, universities are in a unique position to bring in theory, evidence, and partnerships with societal entities to develop "phronesis," or the application of sound principles in complex situations toward some good end. Sustainability education promotes public knowledge, where advocacy is fostered for the greater good. Increased public calls for university accountability to contribute to the public good, beyond offering university degrees or training students for specific jobs, bode well for the role of sustainability education.

While stark disagreements do not appear among the contributors to this book, there are contrasting emphases across the chapters. Perhaps the most provocative emphasis, which has received less attention in scholarly discussions of sustainability education at the post-secondary level, is that of Sharpe (chapter 6), who calls for incorporating more training on the stages of clean technology investment, development, and scaling up. Megginson (chapter 5) also stands out in his arguments about the role of mathematics in tribal college curricula and sustainability education. Whereas Goebel (chapter 12) and Perkins (chapter 11) especially call for sustainability education that requires engaged citizenship (and eschews individualism), Houshmand (chapter 13) emphasizes sustainability education that embraces seeing students as expressing their values through their consumption choices. Whereas Megginson (chapter 5) and Epp (chapter 2) heavily emphasize the importance of developing sustainability programming that is tied to the place where a higher education institution is situated, Dale (chapter 8), Malcom (chapter 9), and Dietz (chapter 7) emphasize developing curriculum around grand challenges, or wicked problems, that transcend borders.

I believe the contributing authors of this book agree that higher education institutions should prepare their students to address the needs of the future. This requires a greater nimbleness than what currently exists, a closer ear to the ground, a greater presence of university experts on decision-making bodies, and meaningful long-term partnerships. Higher education partnerships with civil society, non-governmental organizations, businesses, governments, and other schools connect student and faculty potential to current sustainability challenges. We need more fluid involvement of sustainability practitioners as guest lecturers and co-teachers in our classrooms. Sustainability in higher education can be a model for how to best serve the public good. Most importantly, universities need to teach students more about the successes and experiments underway, to help build the knowledge, confidence, and skills they need to forge a better future for all. I hope this book will inform and bolster your own commitment to be an even stronger part of this.

Note

1. See more on the 26th UN Climate Change Conference of the Parties (COP26), held from October 31 to November 13, 2021, at https://ukcop26.org/.

References

Butcher, Jim. 2007. "Are You Sustainability Literate?" *Spiked*. September 13, 2007. https://www.spiked-online.com/2007/09/13/are-you-sustainability-literate/.

Dawe, Gerald, Rolf Jucker, and Stephen Martin. 2005. *Sustainable Development in Higher Education: Current Practice and Future Developments. A Report to the Higher Education Academy*. November 2005. York (UK): The Higher Education Academy. https://www.heacademy.ac.uk/system/files/sustdevinHEfinalreport.pdf.

Grenier, Éric. 2020. "When It Comes to Leadership, Canada's Political Parties Aren't Getting More Diverse." *CBC News*, June 9, 2020. https://www.cbc.ca/news/politics/grenier-leadership-diversity-1.5603626.

Jones, Paula, David Selby, and Stephen Sterling, eds. 2010. *Sustainability Education: Perspectives and Practice across Higher Education*. New York: Earthscan.

Lozano, Rodrigo. 2010. "Diffusion of Sustainable Development in Universities' Curricula: An Empirical Example from Cardiff University." *Journal of Cleaner Production* 18, no. 7 (May): 637–44. https://doi.org/10.1016/j.jclepro.2009.07.005.

Maslen, Geoff. 2012. "Worldwide Student Numbers Forecast to Double by 2025." *University World News*, February 19, 2012. https://www.universityworldnews.com/post.php?story=20120216105739999.

Max-Neef, Manfred A. 2005. "Foundations of Transdisciplinarity." *Ecological Economics* 53, no. 1 (April): 5–16. https://doi.org/10.1016/j.ecolecon.2005.01.014.

Nicolaides, Angelo. 2006. "The Implementation of Environmental Management towards Sustainable Universities and Education for Sustainable Development as an Ethical Imperative." *International Journal of Sustainability in Higher Education* 7, no. 4 (October): 414–24. https://doi.org/10.1108/14676370610702217.

Peet, D.-J., Karel F. Mulder, and Arianne Bijma. 2004. "Integrating SD into Engineering Courses at the Delft University of Technology: The Individual Interaction Method." *International Journal of Sustainability in Higher Education* 5, no. 3 (September): 278–88. https://doi.org/10.1108/14676370410546420.

Serai, Esha. 2020. "Record Numbers of Millennials Run for Public Office." *VOA News*, November 2, 2020. https://www.voanews.com/2020-usa-votes/record-numbers-millennials-run-public-office.

Thomas, Ian. 2004. "Sustainability in Tertiary Curricula: What Is Stopping It Happening?" *International Journal of Sustainability in Higher Education* 5, no. 1 (March): 33–47. https://doi.org/10.1108/14676370410517387.

ULSF (University Leaders for a Sustainable Future). 2015. "Report and Declaration of the PresidentsConference(1990)."http://ulsf.org/report-and-declaration-of-the-presidents-conference-1990/.

Wiek, Arnim, Lauren Withycombe, and Charles L. Redman. 2011. "Key Competencies in Sustainability: A Reference Framework for Academic Program Development." *Sustainability Science* 6, no. 2 (July): 203–18. https://doi.org/10.1007/s11625-011-0132-6.

Contributors

APRYL BERGSTROM is a sessional instructor at the University of Alberta. She has taught undergraduate courses in environmental sociology and the history and fundamentals of environmental protection and conservation and is currently working on developing a course on the United Nation's Sustainable Development Goals. Her Master of Science degree in rural and environmental sociology at the University of Alberta focused on sustainability courses in higher education. Prior to her graduate work, she completed undergraduate degrees in environmental and conservation sciences, philosophy, and physics. As a research assistant in the University of Alberta's provost's office, she assisted with planning and administering academic sustainability initiatives.

CHRISTOPHER G. BOONE is dean of the College of Global Futures and a professor in the School of Sustainability at Arizona State University. His research contributes to ongoing debates in sustainable urbanization, environmental justice, vulnerability, global environmental change, and innovation in higher education. At ASU, he has taught classes on sustainable urbanization, urban and environmental health, principles and methods of sustainability, environmental justice, interdisciplinary methods for socio-ecological research, urbanization, biodiversity, innovation, and sustainable design (Innovation Space). He earned his PHD in geography (1994) from the University of Toronto and was a post-doctoral fellow in the School of Environment at McGill University.

ANN DALE, professor and director of the School of Environment and Sustainability, Royal Roads University, held her university's first Canada Research Chair in sustainable community development (2004–14), and is a Trudeau Alumna and a Fellow of the World Academy of Art and Sciences. She received the CUFA Paz Buttedahl Distinguished Career Academic Award (2014); the Canada Council for the Arts, Molson Prize for the Social Sciences (2013); and the 2009 Bissett Award for Distinctive Contributions to the Public Sector. Her research focuses on climate pollution, governance, research curation, sustainable community development, social capital, and agency.

THOMAS DIETZ is a University Distinguished Professor, and a professor of environmental science and policy and sociology at Michigan State University. He is active in the animal studies graduate specialization, was founding director of the Environmental Science and Policy Program, and was co-director of the Great Lakes Integrated Sciences and Assessments Center. He is a fellow of the American Association for the Advancement of Science and has received the Sustainability Science Award of the Ecological Society of America. His research focuses on environmental values, macro-comparative human ecology, and the relationship between science and values in democracy.

ROGER EPP is a professor of political science at the University of Alberta, where he has held several senior leadership positions. He has served as founding dean of the university's Augustana Campus in Camrose (2004–2011), as deputy provost, and as director of UAlberta North. He is author of *We Are All Treaty People: Prairie Essays* (2008), co-editor of *Roads Taken: The Professorial Life, Scholarship in Place, and the Public Good* (2014), and author of *Only Leave a Trace: Meditations* (2017). In 2013, he served as an honorary witness at a hearing of the Truth and Reconciliation Commission of Canada.

ALLISON F.W. GOEBEL is a professor of environmental studies at Queen's University, cross-appointed to the departments of gender studies, global development studies, and sociology. She is the author of *Gender and Land Reform. The Zimbabwean Experience* (2005) and *On*

Their Own: Women, Urbanization and the Right to the City in South Africa (2015), and of articles relating to land, gender, livelihoods, urban housing issues, health, and environments in southern Africa. She has taught an introductory course, Environment and Sustainability, for more than ten years and is experimenting with new pedagogies in sustainability education, focusing on interdisciplinarity, global perspectives, and sustainability citizenship.

KOUROSH HOUSHMAND is a behavioural data specialist at McKinsey & Company. He is a recipient of Canada's Top 30 Under 30 sustainability leaders award and the "Climate Hero" award by Canadian Geographic and Shell Canada, and a Rhodes Scholarship finalist. In 2013, he founded an international non-profit organization called Solar for Life, with volunteers across Canada, Kenya, Myanmar, and South Africa. He received a Master of Science degree at Columbia University, where he was awarded the Walter H. Diamond and Dorothy B. Diamond International Business Fellowship. He earned his second master's degree in economics at Peking University, where he was a Yenching Scholar.

ROBERT H. JONES is a professor of forestry and environmental conservation, and executive vice president for academic affairs and provost at Clemson University. He has published sixty-two papers on forest and soil ecology in ecology and forestry related journals. He has directed eighteen graduate theses and obtained $4.6 million in extramural grant funding as a principal or co-principal investigator. He served on EPA and NSF review panels and on the editorial boards for three journals. Jones has taught study abroad, undergraduate, and graduate courses in ecology, and has received five awards for teaching.

NAOMI KROGMAN is the dean of the Faculty of Environment at Simon Fraser University. Prior to this, she was a professor of environmental sociology in the Department of Resource Economics and Environmental Sociology at the University of Alberta for 22 years. From 2012 to 2016, she served as the director of sustainability scholarship and education out of the provost's office, where she developed an undergraduate Certificate in Sustainability, a graduate student Sustainability Scholars program, and an academic plan for sustainability. From 2016 to 2019, she was an associate dean of graduate studies. Her current work focuses on approaches to sustainability in higher education.

SHIRLEY M. MALCOM is senior advisor and director of SEA Change at AAAS. She is a trustee of Caltech, regent of Morgan State University, member of the SUNY Research Council, and former member of the National Science Board, and she served on President Clinton's Committee of Advisors on Science and Technology. She holds seventeen honorary degrees, and serves on the boards of the Heinz Endowments, Public Agenda and the National Math-Science Initiative. She is co-chair of the Gender Advisory Board of the UN Commission on S&T for Development and Gender InSITE, a global campaign to help improve the lives and status of girls and women.

ROBERT E. MEGGINSON is an Arthur F. Thurnau Professor of Mathematics at the University of Michigan, where he has been since 1992, except when on leave to serve as deputy director of the Mathematical Sciences Research Institute (2002–2004). A Fellow of the American Association for the Advancement of Science and the American Mathematical Society and member of the National Science Foundation's CEOSE diversity advisory committee, his recognitions include the US Presidential Award for Excellence in Science, Mathematics, and Engineering Mentoring (1997), the Ely S. Parker Award of the American Indian Science and Engineering Society (1999), and the 2019 SACNAS Distinguished Mentor Award. He was portrayed in *100 Native Americans Who Shaped American History* (2002).

PATRICIA E. (ELLIE) PERKINS is a professor in the Faculty of Environmental and Urban Change, York University, Toronto, where she teaches ecological economics, community economic development, climate change policy, and critical interdisciplinary research design. Her research focuses on feminist ecological economics, climate justice, commons, and participatory governance. In addition to many journal articles and book chapters, she is the editor of *Local Activism for Global Climate Justice: The Great Lakes Watershed* (2019). She directs a Queen Elizabeth Scholars project with partners in Brazil, Chile, South Africa,

Cameroon, Kenya, Mozambique, and Ghana (qesclimatejustice.info.yorku.ca).

VICKY J. SHARPE is a corporate director with expertise in integrating environmental, social, and governance (ESG) criteria into business to maximize holistic returns. As founding president and CEO of Sustainable Development Technology Canada, she built an internationally renowned cleantech fund of 280+ companies and mobilized over $5 billion from public and private sources. A founding director of the Capital Markets Regulatory Authority, Alberta Enterprise Corporation, and Carbon Management Canada Research Institutes, she serves on the boards of EfficiencyOne, The Verschuren Centre, and QUEST (Quality Urban Energy Systems for Tomorrow) and sits on the advisory boards of EnerTech Capital and Mercer's global Sustainable Opportunities Fund.

TODDI A. STEELMAN is the Stanback Dean of the Nicholas School of the Environment at Duke University. Prior to this, she served for five years as executive director of and professor at the School of Environment and Sustainability at the University of Saskatchewan. Her research has been funded by the National Science Foundation and Canadian Tri Agencies and by federal and state agencies. She has authored four books and published extensively in peer-reviewed journals. She is best known as a wildfire expert and has brought her expertise to venues including the Royal Society (UK) and National Academy of Sciences (US).

Other Titles from University of Alberta Press

Making Wonderful
Ideological Roots of Our Eco-Catastrophe
MARTIN M. TWEEDALE
Documents how the West came to have an ideology that has promoted environmentally destructive economic expansion.

Troubling Truth and Reconciliation in Canadian Education
Critical Perspectives
Edited by SANDRA D. STYRES & ARLO KEMPF
Offers a series of critical perspectives concerning reconciliation and reconciliatory efforts between Canadian and Indigenous peoples in the field of education.

Dissonant Methods
Undoing Discipline in the Humanities Classroom
Edited by ADA S. JAARSMA & KIT DOBSON
An innovative collection that urges instructors to make the humanities classroom a space for resistance.

More information at uap.ualberta.ca